# LAW AND POLICY OF SUBSTANTIAL OWNERSHIP AND EFFECTIVE CONTROL OF AIRLINES

# Law and Policy of Substantial Ownership and Effective Control of Airlines

## Prospects for Change

ISABELLE LELIEUR
*UCCEGA – Les Aéroports Français*
*Head of the Legal Department*

Routledge
Taylor & Francis Group

LONDON AND NEW YORK

First published 2003 by Ashgate Publishing

2 Park Square, Milton Park, Abingdon, Oxon OX14 4RN
711 Third Avenue, New York, NY 10017, USA

*Routledge is an imprint of the Taylor & Francis Group, an informa business*

First issued in paperback 2016

**British Library Cataloguing in Publication Data**
Lelieur, Isabelle
 Law and policy of substantial ownership and effective
 control of airlines : prospects for change
 1.Airlines-Ownership 2.Aeronautics, Commercial - Law and
 legislation
 I. Title
 387.7'06

**Library of Congress Cataloging-in-Publication Data**
Lelieur, Isabelle, 1975-
 Law and policy of substantial ownership and effective control of airlines : prospects for
 change / Isabelle Lelieur.
  p. cm.
 Includes bibliographical references and index.
 ISBN 978-0-7546-35481-2 (alk.paper)
  1.Airlines--Foreign ownership. 2. Airlines--State supervision. 3.Aeronautics,
 Commercial--Law and legislation. I. Title.

K4095.L45 2003
387.7'1–dc21

2003043028

Transfered to Digital Printing in 2010


ISBN 978-0-7546-3548-2 (hbk)
ISBN 978-1-138-26434-2 (pbk)

# Contents

# Preface

The airline industry is an odd one, in a number of interrelated ways. For newly independent nations, a national airline is often rated as high a priority as a national currency or an army. It shows that the nation can take care of its own affairs. By 'showing the flag' in foreign airports, the State demonstrates a mix of adulthood, sovereignty and national identity. Of course, a national airline also serves other, more mundane, interests such as trade and tourism, and it may provide commercial and technical jobs, bring in (tax) revenues and produce other direct and indirect benefits to the national economy.

But unlike the hotel or telecom business of a country, a national airline evokes strong emotions which become particularly apparent when the airline's survival is, or appears to be, at stake. When, in the 1970s, the US authorities, on the request of ailing PanAm, demanded that KLM Royal Dutch Airlines halve the number of its flights between Amsterdam and New York, this produced such a nationwide public outcry in the Netherlands, allegedly the US's biggest foreign investor and staunchest NATO ally, that Henry Kissinger and his Netherlands counterpart were forced to make a hasty compromise deal lest the outpouring of nationalist emotion seriously threaten diplomatic relations between the two nations.

Aviation relations form part of bilateral diplomatic relations, not only as a result of the above aspects, but also for legal and practical reasons. Airline operations into foreign countries depend on bilateral air transport agreements, concluded by the governments of the States concerned. Although standardized to a large extent, these agreements may involve lengthy and often thorny negotiations, during which other bilateral interests of a commercial, political or other nature may be brought into the picture. Thus, air transport agreements, and the commercial opportunities the respective airlines derive therefrom, are often seen as part of an overall bilateral balance, to be safeguarded by the governments concerned.

Moreover, with the great majority of airlines state-owned, the above explains to some extent why States traditionally insisted on keeping their airlines in national hands and why all bilateral air transport agreements contained the so-called nationality clause which requires that the airlines of the two countries concerned are 'substantially owned and effectively controlled' by (nationals of) the countries designating the airlines.

This system has been in force since 1944. In the meantime, global aviation has changed in many ways. Following the US's example, more and more airlines were turned into private hands, subsidization was progressively reduced or outlawed (EU) and bilateral restrictions on market access were reduced or removed altogether (open skies agreements, EU internal market). Airlines responded to the resulting sharp increase in worldwide competition by cost cutting, revenue-enhancing measures or by joining forces.

Airlines of the same nationality often engaged in mergers and acquisitions, but, due to the above-mentioned ownership and control restrictions, this was not a path airlines of different nationality could follow. Thus, they engaged in various forms of short and long-term cooperation, in the commercial, operational and technical field. And some airlines created alliances, in particular with the established US carriers and with the blessing of both the US and EU competition authorities. These 'strategic alliances' combine all possible forms of cooperation into something which, in some cases, gets pretty close to a merger. But real international airline mergers have not emerged so far.

As we observed above, the airline industry, compared to other industries, is an odd one. Ms. Isabelle Lelieur, who holds a Master's Degree from McGill University's Institute of Air and Space Law, Montreal, has studied all major aspects of this industry, focusing in her book on the question of ownership and control and, more in particular, on the background of, and rationale for, this bilateral clause. She challenges the continued validity and viability of this restrictive provision in an era of globalization and liberalization and offers convincing arguments for jettisoning the legal, economic and security justifications used so far. Her analysis of the fruits which the industry could reap from liberalized cross-border investments and from foreign take-overs is definitely food for thought, as is her description of the steps to be taken on a national, regional and global level. Ms. Lelieur is firmly in favour of a soonest abolition of this typical airline oddity, the bilateral ownership and control clause, which prevents the airline industry from engaging in activities which are totally accepted in virtually all other international industries. But she is not blind to the various roadblocks that continue to stand in the way of real progress in this regard.

EU–US aviation negotiations, brought closer by the recent judgment of the European Court of Justice in the 'open skies' cases, are often seen as one of the more promising roads to reaching this goal of freedom to invest and merge internationally. That may be so, but the recent downfall of United Airlines may have brought a keener sense of urgency to the US domestic discussion on the pros and cons of the nationality requirement in both domestic law and bilateral agreements. This discussion includes the question of 'which national interests can only be served by an airline in national hands and are therefore threatened when the airline passes into foreign hands'. That question is a crucial one deserving soonest attention, not only on the part of the US but also on the part of any other country whose airline may need a foreign investor to survive in the international market place. The tragic fate of the above-mentioned US mega-carrier may thus be a blessing in disguise for the US and international airline industry.

I recommend this well-researched and amply documented book to the interested reader. With it, Isabelle Lelieur has contributed valuable material and food for thought to the highly topical discussion on the question of airline ownership and control and its impact on the normalization of the global airline industry.

Peter van Fenema
January 2003

# Acknowledgements

I would like to express my sincere thanks to the people who contributed to the accomplishment of the present book by providing me with advice, help and support.

First and foremost, I am indebted to Dr. H.P. van Fenema (Institute of Air and Space Law, McGill University), for his constant encouragement and guidance. His precious advice and assistance have made possible the publication of the book.

I also wish to extend my gratitude to my supervisor, Professor M. Milde (Institute of Air and Space Law, McGill University), as well as to Dr. R.I.R. Abeyratne (Air Transport Bureau, International Civil Aviation Organization), who gave me access to innumerable documents and sources, who provided me with helpful comments, and who generously shared their expertise and experience in the air transport sector.

I would also like to acknowledge Mr. Y. Wang (Economic Policy Section, International Civil Aviation Organization), Mr. L. van Hasselt (Head of Unit, Directorate General for Energy and Transport, European Commission), Mr. W. Hubner (Head of the Division of Transport, Organization of Economic Cooperation and Development), Mr. P. Latrille (Economic Affairs Officer, World Trade Organization), Mr. R. Janda (Professor, Faculty of Law, McGill University), Mr. D. David (Deputy Director of the International Affairs, Directorate General for Civil Aviation, France), Mrs. E. Lacaze (Representative of Multilateral Affairs outside Europe, Directorate General for Civil Aviation, France), Mr. M. Gorog (Director of the International Relations, Air France Cargo), Mr. A. Camus (Direction of International Affairs, Air France) and Mr. F. Gagey (Financial Director, Air France) who sacrificed their time to give me invaluable first-hand information.

Finally, I acknowledge my parents, Geneviève and Christophe Lelieur, for their unfailing emotional support. This book is dedicated to them.

# List of Abbreviations

| | |
|---|---|
| A.A.S.L. | Annals of Air and Space Law |
| ABA | American Bar Association |
| AEA | Association of the European Airlines |
| Air & Space L. | Air and Space Law |
| Air & Space Law. | Air and Space Lawyer |
| Airline Bus. | Airline Business |
| Airline Fin. News | Airline Finance News |
| ALPA | Airline Pilot's Association |
| Am. U. L. Rev. | American University Law Review |
| APEC | Asia Pacific Economic Cooperation |
| ASEAN | Association of Southeast Asian Nations |
| ATRP | Air Transport Regulation Panel |
| BA | British Airways |
| BWIA | British West Indies Airways |
| CAAC | Civil Aviation Administration of China |
| CAB | Civil Aeronautics Board |
| CARICOM | Caribbean Economic Community |
| Case W. Res. J. Int'l L. | Case Western Reserve Journal of International Law |
| CRAF | Civil Reserve Air Fleet |
| CRSs | Computerized Reservation Systems |
| CTC | Canadian Transport Commission |
| DOD | Department Of Defense |
| DOJ | Department Of Justice |
| DOT | Department Of Transportation |
| Duke L.J. | Duke Law Journal |
| EASA | European Aviation Safety Agency |
| ECAC | European Civil Aviation Conference |
| ECJ | European Court of Justice |
| EEA | European Economic Area |
| Emory Int'l L. Rev. | Emory International Law Review |
| FAA | Federal Aviation Authority |
| FDI | Foreign Direct Investment |
| FFP | Frequent-Flier Programs |
| Fordham Int'l L. J. | Fordham International Law Journal |
| GAO | General Accounting Office |
| GATT | General Agreement on Tariffs and Trade |
| GATS | General Agreement on Trade in Services |
| Geo. Wash. J. Int'l L. & Econ. | George Washington Journal of International Law and Economics |

| | |
|---|---|
| Harv. J. on Legis. | Harvard Journal on Legislation |
| IAEAA | International Antitrust Enforcement Act |
| IASTA | International Air Services Transit Agreement |
| IATA | International Air Transport Association |
| ICAO | International Civil Aviation Organization |
| ICAO J. | ICAO Journal |
| IFFAS | International Financial Facility for Aviation Safety |
| Inv. Dealers' Dig. | Investment Dealers Digest |
| ITF | International Transport Workers' Federation |
| JAA | Joint Aviation Authorities |
| J. Air L. & Com. | Journal of Air Law and Commerce |
| J. L. & Com. | Journal of Law and Commerce |
| J. Rec. (Okla. City) | Journal Record of Oklahoma City |
| L. & Econ. R. | Law and Economic Review |
| Marad | US Maritime Administration |
| M&A | Merger and Acquisition |
| MFN | Most Favored Nation |
| NAFTA | North American Free Trade Agreement. |
| NTA | National Transportation Agency |
| OECD | Organization for Economic Co-operation and Development |
| Orange County Reg. | Orange County Register |
| Public Contract L. J. | Public Contract Law Journal |
| SARPs | Standards And Recommended Practices |
| SAS | Scandinavian Airlines System |
| SIA | Singapore International Airlines |
| Suffolk Transnat'l L. Rev. | Suffolk Transnational Law Review |
| Syracuse L. R. | Syracuse Law Review |
| TCAA | Transatlantic Common Aviation Area |
| Transp. L. J. | Transportation Law Journal |
| TWA | Trans World Airlines |
| U. Bus. Miami L. J. | University of Miami Business Law Journal |
| USOAP | Universal Security Oversight Audit Programme |
| U. Haw. L. Rev. | University of Hawaii Law Review |
| US J. Com. | US Journal of Commerce |
| Wall St. J. | Wall Street Journal |
| WTO | World Trade Organization |

# Introduction

## 1. Introduction

Civil aviation is a commercial activity and, as such, is embroiled in the process of globalization currently affecting business pursuits worldwide. International air transport is already one of the world's largest industries; nevertheless, for the industry to flourish in the 21$^{st}$ century, it will require a more liberalized legal and economic framework tailored to the global marketplace. At present, the main impediment to establishing this framework is the traditional requirement that airlines must be *substantially owned* and *effectively controlled* by nationals of the State to which such airlines are linked through the flags of the State concerned. This notion of 'flag carrier' has been the norm in worldwide aviation policy for more than 50 years,[1] and it is firmly entrenched in national laws, as well as in most bilateral and multilateral agreements. Thus, this book proposes the total abolishment of the national ownership and control restrictions in order to allow air carriers to evolve in a more liberal environment and critically reviews the traditional justifications for these legal restrictions.

## 2. Is the 'Ownership and Control Requirement' Part of International Law?

The 1944 *Convention on International Civil Aviation*,[2] which is arguably the basis for the entire international aviation legal system, does not expressly affirm national ownership requirements for airlines. Still, some authors maintain that the Convention gives implicit approval to these restrictions through its recognition of certain principles,[3] such as sovereignty of the State over the airspace above its territory (Article 1),[4] preferential treatment by a

---

[1] P.P.C. Haanappel, 'Airline Ownership and Control and Some Related Matters' (2001) 26-2 Air & Space L. 90 at 90 [hereinafter Haanappel 'Airline Ownership and Control'].
[2] *Convention on International Civil Aviation*, 7 December 1944, 15 U.N.T.S. 295, ICAO Doc. 7300/6 [hereinafter the *Chicago Convention*].
[3] K. Bohmann, 'The Ownership and Control Requirement in U.S. and European Union Air Law and U.S. Maritime Law – Policy; Consideration; Comparison' (2001) 66 J. Air L. & Com. 689 at 692.
[4] Article 1 of the *Chicago Convention* states that '[t]he contracting States recognize that every State has complete and exclusive sovereignty over the airspace above its territory'.

State *vis-à-vis* its national carrier (Article 7),[5] and bilateral negotiations of traffic rights between States (Article 6).[6] However, taken as a whole, the *Chicago Convention* is, at best, neutral with regard to ownership criteria, since it also expressly permits operations involving joint and coordinated efforts among airlines to provide international service (Article 77).

Indeed, the only international agreement that actually addresses the issue of airline ownership restrictions is the *International Air Services Transit Agreement.*[7] Article 1, Section 5, provides that:

> Each Contracting State reserves the right to withhold or revoke a certificate or permit to an air transport enterprise of another State in any case where it is not satisfied that substantial ownership and effective control are vested in nationals of a Contracting State, or in case of failure of such air transport enterprise to comply with the laws of the State over which it operates, or to perform its obligations under this Agreement.[8]

In the period since the 1950s, however, as international civil aviation grew in breadth and scope, States increasingly used the right to withhold or revoke a foreign airline's certificate or permit to operate in their *national* airspace, as a means of regulating *international* air transport. Specifically, States limited foreign investment in their 'flag carriers' by requiring a 'nationality clause' based on Article 1(5) of the IASTA in each of their bilateral agreements on air traffic rights. In this way, airlines from non-contracting countries could not benefit from a bilateral exchange of traffic rights,[9] since only the carriers designated by the contracting parties (and meeting the bilateral agreement's nationality requirements) acquired traffic rights and, *ergo*, had access to the international routes covered by the

---

[5] Article 7 of the *Chicago Convention* states that '[e]ach Contracting State shall have the right to refuse permission to the aircraft of other contracting States to take on in its territory passengers, mail and cargo carried for remuneration or hire and destined for another point within its territory (...)'.

[6] Article 6 of the *Chicago Convention* states that '[n]o scheduled international air service may be operated over or into the territory of a contracting State, except with a special permission or other authorization of that State, and in accordance with the terms of such a permission or authorization'.

[7] *International Air Services Transit Agreement*, 7 December 1944, 84 U.N.T.S. 389, 394, ICAO Doc. 7500, also reproduced in ICAO Doc. 9587. The *International Air Services Transit Agreement* has been ratified, as of 4 October 2000, by 118 States, 17 of which have ratified during the last five years [hereinafter *IASTA*]. The *International Air Transport Agreement* (7 December 1944, 171 U.N.T.S. 387), in its Article 1 § 6, addresses the same issue; however, as only very few States ratified the Agreement (12 States), it is of limited importance.

[8] *IASTA, ibid.* Article 1, Section 5.

[9] Bohmann, *supra* note 3 at 694.

agreement. The first such agreement was signed between the United States and the United Kingdom in 1946 (*Bermuda I*);[10] it thereafter became the model upon which virtually all other bilateral agreements between States were based. As airline ownership restrictions became more pervasive in the negotiation of air traffic rights, they also increasingly found their way into States' domestic laws.[11] Thus, the requirement for 'national ownership and control' of airlines is not mandated by international law, but rather was born out of State practice over the last 50 years.

## 3. What does 'Substantial Ownership' and 'Effective Control' of an Airline Mean?

None of the international treaties related to international air activity (*Chicago Convention, IASTA, and Air Transport Agreement*) defines the terms 'substantial ownership' and 'effective control,' and there are no universally accepted definitions for these terms.

### 3.1 Substantial Ownership

Within the international community at large, ownership of an airline is generally understood to mean ownership of voting shares of the airline stock, and 'substantial ownership' usually equates to owning more than 50 per cent of the voting shares,[12] regardless of whether the shareholder is a public or a private entity. Thus, 'majority ownership is substantial.'[13] In the past, governments frequently held a majority stake in their national carriers and, therefore, foreign ownership was not a concern. Today, however, with the

---

[10] *Agreement Between the Government of the United States of America and the Government of the United Kingdom Related to Air Services Between their Respective Territories*, 11 February 1946, US/UK, 60 Stat. 1499 [hereinafter *Bermuda I*]. The nationality clause is stated in Article 6 of the Appendix to Bermuda 1.

[11] For instance, in the United States, the *Civil Aeronautics Act* of 1938 (*Civil Aeronautics Act*, Pub. L. No. 75-706, 52 Stat. 973 (1938)) required that 75 per cent of a US carrier's voting equity remains in US hands, whereas the *Air Commerce Act* of 1926 (*Air Commerce Act*, Pub. L. No. 69-254, SS 1-14, 44 Stat. 568 (1926)) had only required 51 per cent of US voting equity in US carrier. Notably, under current US law there is the additional requirement that a US carrier be a US citizen. For further analysis of the *Federal Aviation Act* of 1958, see Part 1, Chap. 2, para. 1., below.

[12] The European Commission has interpreted the EC ownership of airlines in its decision on the Swissair/Sabena case, see EU, *Commission Decision No. 95/404/EC on a Procedure Relating to the Application of the Council Regulation 2407/92 (Swissair/Sabena)*, O.J. (1995) L 239/19.

[13] IATA, Government and Industry Affairs Department, *Report of the Ownership & Control Think Tank World Aviation Regulatory Monitor*, IATA doc. prepared by H.P. van Fenema (7 September 2000) at 13 [hereinafter IATA doc.].

wave of privatization passing through the airline industry, the majority of voting shares in most airlines is now in private hands, with ownership often spread among national and foreign shareholders. Under these circumstances, the percentage of the ownership of voting shares in an airline is not necessarily determinative in establishing 'substantial ownership.' For example, if 60 per cent of an airline's voting shares are spread among different foreign airlines, and nationals hold the remaining 40 per cent, the airline may still remain 'substantially owned' by nationals, as the latter represent the most important owners.[14]

## 3.2 Effective Control

The question of 'effective control' is subtler and requires a deeper analysis, since it has nothing to do with numbers but rather with the question of who actually controls the airline. Control over a corporation is commonly understood as the power to direct its internal and external policy. Such power is normally vested in the board of directors or executive officers of the corporation, as opposed to the shareholders. As van Fenema explains, '[t]o be the national majority shareholder is one thing, but the right to "hire and fire", to set the corporate goals, to take major decisions effecting the future of the company, if such powers reside in other than national hands, will create serious doubts about the nationality of the airline'.[15]

In the United States, the *Securities Exchange Act* of 1934 defines 'control' as follows:

> The term 'control' (including the terms 'controlling,' 'controlled by' and 'under common control with') means the possession, direct or indirect, of the power to direct or cause the direction of the management and policies of a person, whether through the ownership of voting securities, by contract, or otherwise.[16]

Notwithstanding this general definition, US aviation law fails to clarify the notion of what constitutes 'effective control' of airline. Moreover, aviation regulators in the United States, beginning with the US Civil Aeronautics Board (CAB), and thereafter with the US Department of Transportation (DOT), decided the issue of 'effective control' without defining it. Consequently, the meaning of 'effective control' has been clouded by its many different interpretations.

---

[14] *Ibid.* at 13-14.
[15] *Ibid.* at 15.
[16] The *Securities Exchange Act*, 17 C.F.R. § 240.12b-2 (1988 & Supp. 1995).

In practice, the US has established a 'control test', which has been widely applied, first by the CAB and then by the DOT.[17] In its 1989 KLM/Northwest decision, the DOT explained that its analysis of effective control 'has always been on a case-by-case basis, as there are myriad potential avenues of control (...).'[18] Most recently, 'the DOT has looked closely not only at how investment in the US airline industry allows foreign control, but also at how personal relationships between US citizens and foreign purchasers may provide a more subtle method of influence.'[19] In this way, the DOT has made a distinction between two types of foreign control: financial control through equity ownership and control through personal relationships.[20] In essence, the 'control test' is a subjective determination by the DOT of who is actually controlling the airline. Consequently, satisfaction of the US statutory requirements for 'substantial ownership' is not sufficient to determine whether an airline is a domestic (US) airline – the airline must also be deemed to be 'effectively controlled' by US citizens.[21]

In contrast to the situation in the US, the European Union has established a definition of 'effective control' in Council Regulation No. 2407/92 of 23 July 1992,[22] which is used to determine if a national carrier from any EU Member States can be considered a community carrier. Article 2 (g) states:

'Effective control' means a relationship constituted by rights, contracts or any other means which, either separately or jointly and having regard to the considerations of fact or law involved, confer the possibility of directly or indirectly exercising a decisive influence on an undertaking, in particular by:

(a)  the right to use all or part of the assets of an undertaking;
(b)  rights or contracts which confer a decisive influence on the composition, voting or decisions of the bodies of an undertaking or otherwise confer a decisive influence on the running of the business of the undertaking.[23]

---

[17] A. Edwards, 'Foreign Investment in the U.S. Airline Industry: Friend or Foe?' (1995) 9 Emory Int'l L.R. 595 at 627-628, note 205.

[18] Department of Transportation, *Order in the Matter of the Acquisition of Northwest Airlines by Wings Holdings, Inc.*, DOT Order 89-9-29, Docket No. 46371 (29 September 1989) [hereinafter DOT Order 89-9-29].

[19] Edwards, *supra* note 17 at 628.

[20] Bohmann, *supra* note 3 at 698.

[21] J.D. Brown, 'Foreign Investment in U.S. Airlines: What Limits should be Placed on Foreign Ownership of U.S. Carriers?' (1990) 41 Syracuse L.R. 1269 at 1275-76.

[22] EU, *Council Regulation 2407/92 on Licensing of Air Carriers*, [1992] O.J. L. 240/1 [hereinafter *Council Regulation 2407/92*].

[23] *Ibid.* Article 2(g).

The European Commission applied this definition in its 1995 assessment of whether Swissair controlled the European carrier Sabena. The Commission analyzed the composition and powers of the Swiss–Belgian management board, the procedure for the appointment of its chairman, the powers of the CEO and of the Belgian shareholders, and the extent of the Swiss veto rights.[24] The European legislation thus, on its face, appears to be more objective than the US 'control test,' at least in so far as it expressly defines 'effective control.' Moreover, it should be noted that the principle of primacy of EU laws over the national laws of Member States makes *Council Regulation 2407/92* applicable to every EU Member State; consequently, analysis of 'effective control' of European carriers is simplified to the extent that national laws no longer have to be taken into account.[25]

In the end, the 'substantial ownership and effective control' requirement in both the US and the EU is intentionally ambiguous, so that it can be applied in a way that supports the policies that the State wants to promote in light of the given economic and political situation. Moreover, there are three additional factors that contribute to the ambiguity in this area. First, the relevant definitions in national legislation are normally distinct from the definitions used in designation provisions in bilateral agreements and, consequently, 'there may be a partial overlap between these two sets of regulations.'[26] Second, criteria are interpreted on a case-by-case basis and therefore, national interpretations go even beyond the scope of the law.[27] Third, as was previously mentioned, ownership and control are two independent criteria, so an airline can own the majority of the voting shares of another airline without controlling it (thus, a more in-depth analysis must be done in order to answer questions such as assessing who owns the means of production, what are the voting conditions between the shareholders, etc.). Likewise, an airline may own a minority of shares in a foreign airline and yet be deemed to control it.

---

[24] IATA doc., *supra* note 13 at 16.

[25] For instance, French law provides a specific definition of control, see Loi sur les Sociétés Commerciales No. 66-537, 24 July 1966, Article 355-1 al.1 (amended by the Law No. 85-705, 12 July 1985): 'Une société est considérée comme en contrôlant une autre lorsqu'elle détient directement ou indirectement une fraction du capital lui conférant la majorité des droits de vote dans les assemblées générales de cette société.'

[26] WTO, *Note on Developments in the Air Transport sector Since the Conclusion of the Uruguay Round, Part Five.* WTO Doc. S/C/W/163/Add.4 (2001) 5 [hereinafter WTO doc.1].

[27] For an interpretation of the US federal law by the DOT, see e.g., T.D. Grant, 'Foreign Takeovers of United States Airlines: Free Trade Process, Problems and Progress' (1994) 31 Harv. J. Legis. 63 at 101.

## 4. Why Has the National Ownership and Control Requirement Emerged in the Airline Industry and Does it Still Have a Place in the Current Global Economic Environment?

In the first part of the 20th century, many factors necessitated national ownership restrictions in the nascent airline sector. Although ownership restrictions did not make their way into negotiations for international air traffic rights and bilateral agreements until after 1944,[28] such restrictions were in place in some national laws as early as the 1920s.[29] Such restrictions were a natural by-product of the well-established principle of State's sovereignty over the airspace above its territory. Then, as States undertook to grant international air traffic rights through bilateral agreements, ownership restrictions ensured that the State party to the agreement remained the beneficiary of the authorization by preventing the designated carrier from being owned and/or controlled by a government or nationals of a third country. Apart from this, States also applied the principle of national ownership and control to airlines for more general reasons. First, airlines were seen as symbols of national prestige; indeed, '[c]ivil aviation, associated with rapid progress of technology and continuous changes and innovations, has become a mirror reflecting the general standard of [national] society.'[30] Thus, to attack a national symbol, acts of terrorism were used against the airlines, especially during the sixties.[31] Secondly, a national airline provides a measure of political and economic independence, particularly in the event of a national security threat. In fact, national security was the main justification cited by States for imposing national 'substantial ownership and effective control' restrictions on airlines during the period from the 1920s through the 1940s, a period marked by two world wars. Thereafter, national security remained a primary basis for the imposition of ownership restrictions on airlines due to the Cold War.[32]

In recent decades, the justifications for ownership restrictions on airlines have become increasingly economic. To a global extent, before the 1970s, national protectionism was economically justifiable in most sectors.[33] In the

---

[28] The *Chicago Convention*, *supra* note 2; *IASTA*, *supra* note 7.

[29] For instance, the *US Air Commerce Act* of 1926 codified restrictive rules, requiring air carriers to maintain 51 per cent of voting stock under US citizenship and a 66¾ per cent US citizen contingent on their board of directors, see the *US Air Commerce Act*, *supra* note 11.

[30] J.S. Gertler, 'Nationality of Airlines: Is It a Janus with Two (or More) Faces?' (1994) 19:1 A.A.S.L. 211 at 242.

[31] *Ibid*. at 243.

[32] In the US, laws on foreign ownership of US air carriers have emerged because of the constant threat to national security, see C.G. Alexandrakis, 'Foreign Investment in U.S. Airlines: Restrictive Law is Ripe for Change' (1994) 4 U. Bus. Miami L.J. 71 at 73.

[33] The airline industry started to change mostly after the US airline deregulation of 1978, see the *Airline Deregulation Act*, Pub. L; No. 95-904, § 102(7), (10), 92 Stat. 1705 (codified as amended at 49 USC § 1301-1552 (1982)) [hereinafter the *Airline Deregulation Act*].

event of war or international economic crisis, States could hide behind their sovereignty and still maintain their individual markets. However, today, the international scene has changed, through privatization of more and more public undertakings and liberalization of national and international markets. Globalization, in the sense that companies around the globe are closely linked and increasingly exchange goods and services,[34] prevails on the international scene in virtually all major sectors of the world economy. This new situation has lead to increased competition. The international air transport sector is thus directly concerned by this global economic picture.[35] Airlines have been undergoing privatization for more than 30 years,[36] though this process is far from complete, as many airlines are still owned by their governments.[37] To survive the increased competition, many airlines have had to concentrate,[38] and join forces through alliances, which include e.g. code-sharing, franchising, joint venture equity.[39] Accordingly, because of the increasing economic threats the air carriers have to face, ownership restrictions have become a means of protecting national airlines from foreign competition and maintaining jobs in the domestic airline industry.

The relationship between international trade and investment must also be considered within the context of the present discussion as it is directly related to the issue of foreign investment restrictions on airlines. Despite the remarkable growth in world trade, there are still many restrictive laws relating to foreign direct investment (FDI); hence, FDI has become one of the most controversial areas of international law. The benefits of FDI on the national economies of developing as well as developed countries are essentially three-

---

[34] For an overview on what is globalization, see C.W.L. Hill, *International Business: Competing in the Global Marketplace*, third ed. (Boston: Irwin/McGraw-Hill, 2001) 1-30 [hereinafter Hill 2001].

[35] Although this affirmation has to be nuanced. National security is the main counter-argument that makes the aviation industry a particular industry. 'National security' will be discussed in Part 2, Chap. 3, para. 3, below.

[36] ICAO, *The World of Civil Aviation, 2000 – 2003* (Provisional publication of the Circular 287), ICAO Doc. AT/122 (9 October 2001) [hereinafter ICAO Doc. AT/122]; for the evolution of ownership of the European major airlines from 1979 to 1992, see P.S. Dempsey, 'Competition in the Air: European Union Regulation of Commercial Aviation' (2001) 66 J. Air L. & Com. 979 at 983, note 4 [hereinafter Dempsey 'Competition in the Air'].

[37] D. Knibb, 'Thai Moves Towards Privatisation' *Airline Bus.* (December 2000) 24; C. Baker, 'History Lessons' *Airline Bus.* (December 2000) 74.

[38] ICAO, Working Paper (*World-Wide Air Transport Conference on International Air Transport Regulation: Present & Future*), No. AT Conf/4 – WP 5 (8 August 1994); M. Brenner, 'Airline Deregulation – A Case Book in Public Policy Failure' (1988) 16 Transp. L. J. 179 at 181.

[39] S. Tiwari & W.B. Chik, 'Legal Implications of Airline Cooperation: Some Legal Issues and Consequences Arising from the Rise of Airline Strategic Alliances and Integration in the International Dimension' (2001) J. of Aviation Management of Singapore Aviation Academy 9 at 25.

fold: FDI supplies capital, technology and management resources that would otherwise not be available, FDI increases annual global flow of investment, and FDI allows multinational firms to extend their activities internationally.[40]

The foregoing discussion raises the question of whether the principle of national 'substantial ownership and effective control' of airlines is still legitimate in the current global economic scheme. The proliferation of foreign investments between airlines arguably highlights the need to remove national restrictions. Whereas transnational ownership was a marginal phenomenon before the 1990s, in early 2001, more than 57 carriers reportedly held shares in foreign airlines, and over 160 airlines have foreign equity ownership.[41] Moreover, '[m]any regions are involved [and], [n]ot surprisingly, FDI from developed countries is an important part of the total, complemented by 'north-north' investment flows, 'south-south' investments, and even investments from transition or developing economies in airlines of developed countries.'[42]

Throughout Europe, an airline partnership and merger movement is looming. Within the EU itself, national ownership requirements have been replaced by the EC licensing regulation requirement of 'EC ownership' in the Community; therefore, since 1993,[43] European airlines have been free to cooperate in their equity operations.[44] Dempsey affirms that '[t]oday the EU commercial aviation market is well on its way to becoming a market without state-imposed anti-competitive restrictions. Some experts predicted that liberalization would force unprofitable carriers out of business, into mergers, or into buys-outs.'[45] However, the ten last years of freedom of cooperation have not been really demonstrative in terms of mergers among EU airlines. No European international airline has merged since 1993 despite the propitious European legislation. By merging, the risk for European carriers to lose traffic rights previously granted by third countries is the main reason. Likewise, between the EU and third countries, the movement towards equity cooperation is pretty shy. As stated, 'as the British Airways/American Airlines alliance and the Boeing/McDonnell Douglas merger case both demonstrate, EU commercial aviation players are realizing the importance of banding together in an

---

[40] S.S. Haghighi, *A Proposal for an Agreement on Investment in the Framework of the World Trade Organization* (L.L.M. Thesis, Institute of Comparative Law, McGill University 1999) [unpublished] [footnotes omitted].

[41] WTO doc. 1, *supra* note 26, at 4.

[42] *Ibid.*

[43] EU *Council Regulation 2407/92, supra* note 22.

[44] R. Polley raises an interesting question on this issue 'why more mergers between carriers in different Member States have not occurred following liberalization?' R. Polley explains that national restrictions have been removed inside the Community, 'the main problem, however, is that national ownership requirements persist in bilateral air service agreements between Member States and third countries (...)', see R. Polley, 'Defense Strategies of National Carriers' (2000) 23 Fordham Int'l L.J. 170 at 192.

[45] Dempsey 'Competition in the Air', *supra* note 36 at 984.

increasingly global aviation marketplace.'[46] One of the latest examples is Air France's acquisition of a large stake in Air Afrique, increasing its participation from less than 12 per cent of the capital to 35 per cent of the capital.[47] However, the international air transport industry counts very few examples of such acquisitions. The national ownership restrictions, and certainly not the will of the industry, explain the timid equity cooperation between airlines.[48] Nevertheless, these transactions would undoubtedly increase and perhaps become the norm (greatly increasing growth in the airline industry) if they were not limited by foreign investment restrictions. Thus, the removal of foreign ownership restrictions is advocated to allow air carriers to act according to their will and their needs.

The main objective of this book is to identify those factors that still justify the imposition of national ownership restrictions on airlines and to examine the prospects for change in the current policies and regulatory regimes that support them. This objective will be developed in three parts.

Part I will present, in two chapters, the 'great paradox' of the international airline industry. Specifically, Chapter 1 discusses the current situation of an aviation market that is attempting to address globalization and an increasingly competitive market through international consolidation, yet is restricted by persistent national regulatory constraints. The concept of 'sustainability' of airlines, in the current global world of aviation will also be addressed. Chapter 2 examines the legal regimes on ownership and control of airlines, focusing on the United States, Europe and Canada, with special emphasis placed on the increasing number of deviations and exceptions to the standard of 'substantial ownership and effective control' of airlines – the author maintains that this represents a progressive shift towards a more liberalized market.

Part 2 will explore, in two chapters, the validity of the justifications for national ownership restrictions. Chapter 3 will address the issue of whether there is anything unique to the airline industry that would justify the imposition of restrictions on foreign investments in domestic airlines, and whether the public interest is truly being served by these constraints. Through this analysis, it will be determined whether there are still good reasons for keeping national restrictions on foreign investments in this industry. Chapter 4 will look at the legal and economic consequences of revamping the current regime. The author will outline the benefits of liberalizing airline ownership rules and advocate a total abolition of the restrictions.

---

[46] *Ibid.* at 985.

[47] T. Kouamouo, 'Air Afrique passe sous la tutelle d'Air France' *Le Monde* (17 Août 2001) A12.

[48] For instance, Swissair acquired 42 per cent stake in Portugalia in 1999, see S. Montlake, 'Sair Takes Portugalia Stake' *Airline Bus.* (August 1999) 20.

Part 3 will study the prospects for change through three chapters. Chapter 5 will concentrate on the question of what measures are necessary to facilitate the abolition of foreign investment restrictions, such as the liberalization of traffic rights. Chapter 6 will examine the possibility of eliminating ownership restrictions at the regional level as a precursor to the elimination of restrictions worldwide. Chapter 7 will analyze the proposals of the Organization for Economic Co-operation and Development (OECD), the World Trade Organization (WTO), and the International Civil Aviation Organization (ICAO), regarding the air transport liberalization process and, in particular, the ownership and control issue, as well as the role these international organizations can play in this process.

Finally, in the Conclusion, it is submitted that the aviation industry is mature enough today to benefit from a new regulatory framework, with less legal and economic constraints. The lifting of the foreign ownership restrictions of air carriers would considerably foster the consolidation of the industry on the regional level, and, in the long-term, on the multilateral level. The year 2001, and especially the terrorist attacks of 11 September 2001, has resulted in a loss of confidence in the air transport industry, and therefore, the rationale of the ownership and control restrictions returns now to the concept of national identity. However, surprisingly, States seem to consider that passengers will regain confidence in the industry not by establishing a protectionist policy regarding foreign investments, but rather by opening their airlines to foreign States. Indeed, increased cooperation, through cross-border investments, will undoubtedly contribute to the growth and expansion of the airline industry.[49]

---

[49] The information in this book is up to date as of December 2002.

# PART 1

# THE PARADOX OF THE INTERNATIONAL AIRLINE INDUSTRY:

# RESTRICTIONS ON GLOBALIZATION IN AN INCREASINGLY GLOBAL MARKET

# Introduction to Part 1

Although there has been some loosening of national ownership and control regulations, progress in this regard has been slow, arduous, and limited. On the one hand, most States understand the necessity of relaxing market access and other national restrictions to foster the growth of international air transport, to the benefit of the national and global economy. Accordingly, they have begun liberalizing the air transport sector and enhancing competition by opening their markets to foreign air carriers and agreeing to more and more freedoms of the air with their negotiating partners. On the other hand, not all national restrictions have been removed and those mandating substantial ownership and effective control of airlines by nationals are among the most persistent. Thus, although the industry itself appears to recognize the need for change, the reaction of States has been slow. One reason is the complex processes that often go hand-in-hand with legal reform or the divergent political forces at work within certain States, such as the US, or communities of States, such as the EU. But the main justification of the inactions of the States remains the fear that their national carriers do not survive to the increased competition that liberalization of the ownership restrictions may lead. This is the paradox of the international civil aviation industry today: the 'fundamental contradiction in the airline industry' is between the national ties and the international activities of airlines.[1] However, an increasing number of deviations and exceptions to the ownership and control principle reflects the progressive removal of this rule, thus enabling the airline industry to better face the globalization of air transport.

---

[1] P.P.C. Haanappel, 'Airline Challenges: Mergers, Take-overs, Alliances and Franchises' (1995) 21 A.A.S.L. 179 : 'International airlines are almost invariably national rather than multinational companies, but their activities, by their very nature, cross national boundaries, and do so more rapidly and frequently than any other means of transport. Airline activities are therefore largely international in nature' [hereinafter Haanappel 'Airline Challenges'].

Chapter 1

# Towards an Increasingly Global Open Market

## 1. Introduction

'Globalization of the world economy is not an option we can either embrace or reject, it is already a fact of life.'[1] To achieve air transport globalization,[2] complete liberalization of the sector is required, including the removal of national ownership restrictions; thus, the question is not whether to liberalize, but how. Deregulation of air transport[3] was a major step towards liberalization, but it was only the first

---

[1] Speech given by L. de Palacio, 'Globalization – The Way Forward' (IATA World Transport Summit, Madrid, May 27-29, 2001) [hereinafter de Palacio 2001].

[2] For an overview on what is globalization, see C.W.L. Hill, *International Business: Competing in the Global Marketplace*, third ed. (Boston: Irwin/McGraw-Hill, 2001) 1-30 [hereinafter Hill 2001].

[3] About the 1978 US deregulation, see the *Airline Deregulation Act*, *supra* note 33; the EU liberalization was introduced in three steps through 'packages' of legislation in **1987** (EU, *Council Regulation 3975/87, Laying Down the Procedure for the Application of the Rules on Competition to Undertakings in the Air Transport Sector*, [1987] O.J. L. 374/1; EU, *Council Regulation 3976/87, on the Application of Article 85(3) of the Treaty to Certain Categories of Agreements and Concerted Practices in the Air Transport Sector*, [1987] O.J. L. 374/9; *Council Directive 87/601/EEC, on Fares for Scheduled Air Services between Member States*, [1987] O.J. L. 374/12; and EU, *Council Decision 87/602/EEC on the Sharing of Passenger Capacity between Air Carriers on Scheduled Air Services between Member States and on Access for Air Carriers to Scheduled Air Service Routes between Member States*, [1987] O.J. L. 374/19; **1990** (EU, *Council Decision 87/602/EEC on the Sharing of Passenger Capacity between Air Carriers on Scheduled Air Services between Member States and on Access for Air Carriers to Scheduled Air Service Routes between Member States*, [1987] O.J. L. 374/19; EU, *Council Regulation 2343/90 on Access for Air Carriers to Scheduled Intracommunity Air Service Routes and on the Sharing of Passenger Capacity between Air Carriers on Scheduled Air Services between Member States*, [1990] O.J. L. 217/8; EU, *Council Regulation 2344/90 Amending Regulation 3676/87 on the Application of Article 85(3) of the Treaty to Certain Categories of Agreements and Concerted Practices in the Air Transport Sector*, [1990] O.J. L. 217/15); and in **1992** (EU, *Council Regulation 2407/92, supra* note 22; EU; *Council Regulation 2408/92 on Access for Community Air Carriers to Intracommunity Air Routes*, [1992] O.J. L. 240/8 (corrected in [1992] O.J. L. 15/33); EU, *Council Regulation 2409/92 on Fares and Rates for Air Services*, [1992] O.J. L. 240/15; EU, *Council Regulation 2410/92 Amending Regulation 3975/87 Laying Down the Procedure for the Application of the rules to Competition to Undertakings in the Air Transport Sector*, [1992] O.J. L. 240/18; EU, *Council Regulation 2411/92*

step in the process. Indeed, multilateral negotiations and international regulations are still needed to replace the fifty-year-old bilateral process in place today. Still waiting for a regulatory shift, the survival of a number of air carriers is currently threatened by the new aviation market. Moreover, due to the concentration of airlines brought on by liberalization, continued cooperation between airlines is necessary to gradually break down the national regulatory constraints on ownership and control.

## 2. Consolidation and Multilateralism Required by the New Global Market

### 2.1  A Progressive Shift from Concentration to Consolidation

The meaning of 'market' in aviation terms has changed. It is no longer accurate to talk about markets in terms of limited geographic regions; the aviation market has become global and it can no longer be claimed that competitive practices in one region have little impact on the airlines in other regions.[4] Moreover, globalization will likely be accelerated by the staggering growth of international civil aviation that is expected in the years ahead.[5] As a result, airlines all over the world have started to cooperate to face competition;[6] the industry has been pushed towards concentration, as only few global mega-carriers cover most of the world market today.[7] The current strategy adopted by airlines is to expand in order to achieve economies of scale, global marketing, and a presence in a new market. About 600 all type-agreements among 220 carriers exist.[8] Increased size can in principle be realized through a multitude of arrangements between air carriers, such as mergers, equity investments, global alliances,[9] code-sharing agreements,[10] and franchise

---

*Amending Regulation 3976/87 on the Application of Article 85(3) of the Treaty to Certain Categories of Agreements and Concerted Practices in the Air Transport Sector*, [1992] O.J. L 240/19).

[4] For further discussion on globalization, see the report of the fourth European Air Transport Conference held in Brussels in 1997, H.L. van Traa-Engelman, 'Reports of Conferences: The European Air Transport Conference – Airline Globalization' (1998) 23:1 Air & Space L. 31.

[5] Before the events of 11 September 2001, ICAO foresaw that, in Europe, scheduled passenger traffic was going to grow at rate of 6 per cent in 2001: R.I.R. Abeyratne, 'Emergent Trends in Aviation Competition Laws in Europe and in North America' (2000) 23 World Competition. R. 141 at 141[hereinafter Abeyratne 'Aviation Competition Laws'].

[6] For a comparative assessment of air transport competition in Europe and in North America, see *ibid.* at 145 and 155.

[7] P.S. Dempsey, 'Airlines in Turbulence: Strategies for Survival' (1995) 23:15 Transp. L. J. 15 at 97 [hereinafter Dempsey 'Airlines in Turbulence']; R. Doganis, 'Relaxing Airline Ownership and Investment Rules' (1996) 21 Air & Space L. 267.

[8] WTO, *Note on Developments in the Air Transport Sector Since the Conclusion of the Uruguay Round, Part Five.* WTO Doc. S/C/W/163/Add.4 (2001) 8 [hereinafter WTO doc.1].

[9] *Ibid.* at 13.

[10] *Ibid.* at 11.

agreements.[11] However, carriers most frequently cooperate through 'code-sharing,' whereby airlines place their code on the flights of another carrier, sell the service as their own. In addition, modern alliances increasingly seek to develop synergies, by exploiting common routes, infrastructures and services. Thus, although international mergers are still rare, 'during 1999 there were $150 billion worth of global airline Merger and Acquisition (M&A) transactions, according to Thomson Financial Securities Data'[12] – analysts refer to these transactions as 'virtual mergers'.

Yet, more and more, cross-border acquisitions and even mergers have begun to replace the more traditional alliance agreements as the industry moves from an era of concentration to a period of consolidation, in which dominant airlines seek increased control over and not merely cooperation with their foreign alliance partners.[13] Most recently, consolidation can be seen on the national level in China, where, in September 2000, the Civil Aviation Administration of China (CAAC) demanded drastic consolidation of the country's more than 30 carriers, and by February 2001 a consolidation was in place.[14] Of course, consolidation extends to the international scene as well, but it is limited due to national requirements for carriers to be substantially owned and effectively controlled by national interests. In the face of these impediments, airlines of differing nationalities needing to consolidate are forced to enter into strategic alliances rather than merge, thereby 'keeping their national identities, at least formally, intact.'[15] Accordingly, full international consolidation will not be possible as long as national restrictions on ownership are maintained.

---

[11] For the definition of an airline franchise, see P.P.C. Haanappel, 'Airline Ownership and Control and Some Related Matters' (2001) 26-2 Air & Space L. 90 at 180 [hereinafter Haanappel 'Airline Ownership and Control'].

[12] S. Gawlicki, 'Virtual Mergers: With Traditional Mergers Difficult to Pull off, Airlines Finding Creative Ways to Consolidate' (2000) Inv. Dealers' Dig. (WL 4666779).

[13] IATA, Government and Industry Affairs Department, *Report of the Ownership & Control Think Tank World Aviation Regulatory Monitor*, IATA doc. prepared by H.P. van Fenema (7 September 2000) at 7 [hereinafter IATA doc.].

[14] Under the plan, Air China will take over China Southwest Airlines and CNAC-Zhejiang Airlines; China Eastern will take over China Northwest Airlines and Great Wall Airlines; and China Southern will take over China Northern Airlines, China Xinjiang Airlines and Yunnan Airlines, see N. Ionides, 'China Merger Takes Shape' *Airline Bus.* (February 2001) 26; and the CAAC might make the decision to give up ownership of airlines and strip this CAAC of equity ties to the 10 Chinese airlines, divided into three groups (Air China, China Eastern and China Southern), to allow them to compete more effectively, see N. Ionides, 'China to Loosen Central Ownership' *Airline Bus.* (April 2001) 27; 'China opens up aviation industry' (8 August 2002) CNN, online CNN
http://www.cnn.com/2002/WORLD/asiapcf/east/08/08/china.aviation/ (date accessed: 20 September 2002).

[15] Haanappel, *supra* note 11 at 181.

*2.2 The Need for a Different System: from Bilateralism to Multilateralism*

The regulation of trade in international air transport services involves an elaborate system of bilateral agreements which identify the airlines of the contracting States which may operate the agreed routes, determine the capacity that can be provided by each of those designated airlines, and specify the rights that may be exercised. Under this regime, competition on each route is limited to suppliers designated by the States under air services agreements, and nationality requirements are imposed upon the designated carriers (i.e., the 'nationality clause') – these are the main characteristics of the 'bilaterals.'[16] Many major actors in aviation still extol the bilateral system for its merits,[17] and most of the cooperative agreements between airlines are still primarily bilateral.[18] Furthermore, although multilateralism is increasing at the regional level, bilateralism still reigns supreme at the inter-regional level.[19] Bilateral agreements have become more liberal with the onset of the US Open-Skies agreements[20] – the first such agreement having been signed in 1992 between the US and the Netherlands.[21] However, true economic liberalization of the airline industry will never be achieved through bilateral agreements so long as they remain discriminatory, with their benefits (e.g., liberalization of traffic rights between the partner States) confined to the airlines of the nations that are party to the agreements by virtue of the nationality clause.[22]

Why then is the question of a change of the bilateral system such a significant issue for the future of the air transport sector?[23] First of all, national

---

[16] For an overview of the evolution, the process, and the structure of the bilateral regulation, see ICAO, *Manual on the Regulation of International Air Transport*, ICAO Doc 9626 (1st ed.) (1996).

[17] For the five main advantages of the bilateral system, see R.D. Lehner, 'Protectionism, Prestige, and National Security: The Alliance Against Multilateral Trade in International Air Transport' (1995) 45 Duke L.J. 436 at 446.

[18] Agreements such as joint venture services, code-sharing, aircraft lease, cargo handling, franchise, maintenance.

[19] J.M. Feldman labels as a 'catastrophe' the fact that, despite the worldwide liberalization, 'the bilateral process is moving backward', see J.M. Feldman, 'No Guts, no Glory' *Air Transport World* (1 January 1992) 65.

[20] J.M. Feldman, 'It's Still a Bilateral World' *Air Transport World* (August 1997) 35.

[21] *Agreement Between the United-States of America and the Netherlands Amending the Agreement of April 3, 1957, as Amended and the Protocol of March 31, 1978, as Amended*, 14 October 1992, US-Neth., T.I.A.S. 11976.

[22] 'These 'Open Skies' agreements do not pave the way for responsible globalization, but continue to be bilateral in nature', see de Palacio 2001, *supra* note 1.

[23] In 1949 already, the bilateral system was questioned, mainly because of its discriminatory character. The doctrine raised the following issue: 'The issue is not multilateralism vs. bilateralism *per se*. The question is rather whether a multilateral formula can be found which will mitigate the disadvantages of a bilateral system, without requiring individual states to sacrifice too many of its advantages.' Thus, since the 1930s, many proposals have been made in favor of a creation of a multilateral system, the first one was the proposal for world ownership and operation of air services on trunk routes submitted by a French group in

constraints, like the national ownership and control requirement, stem from the bilateral agreements.[24] The nationality of the air carrier guarantees that each State gets its own share of the market, with no third parties being allowed to benefit from the bilateral exchange of traffic rights and, thus, under this regime, it is impossible for an air carrier to sell a majority of its shares to a foreign carrier. Therefore, the airline industry cannot consolidate if it continues to be limited by the nationality clause. As Lipman noted, 'a system with its fundamental characteristics determined by constraints on ownership of, and investment in, airlines, and controls on market access, capacity and price, is inconsistent with the general industrial trade liberalizing approaches being pursued in other economic sectors.'[25] Indeed, the bilateral system does not encompass the multinational market access required by the new global system, and it can no longer efficiently accommodate the growing globalization of markets, which indicates that 'these agreements are increasingly out of date and ill adapted to the needs of global operators'[26] and therefore '[t]he airline sector must be liberated from its bilateral straitjacket.'[27]

Indeed, the airline industry needs a new negotiating process that will foster competition across a broader range of markets than is available in the current bilateral system, which is a barrier to air transport growth. Multilateralism must be the new norm for negotiations of air traffic rights between States. A shift away from bilateralism will likely be spawned by the consolidation that is now underway. With the creation of regional economic blocs, such as the North American Free Trade Agreement (NAFTA)[28] and the development of the Association of Southeast Asian Nations (ASEAN)[29], allowing the Member States

---

1933. But all the proposals were rejected by the international community, see V. Little, 'Control of International Air Transport' (1949) III International Organization 29.

[24] The first bilateral agreement was *Bermuda 1* concluded in 1946 between the US and the UK (*Bermuda I, supra* note 10). Even though it was replaced in 1976 by the *Agreement Between the Government of the United States of America and the Government of the United Kingdom Related to Air Services Between their Respective Territories*, July 23, 1977, 28 U.S.T. 5367 [Hereinafter *Bermuda II*], *Bermuda 1* served as a model for all the bilaterals concluded afterwards on the issue of airline nationality. For more details about *Bermuda 1*, see K. Bohmann, 'The Ownership and Control Requirement in U.S. and European Union Air Law and U.S. Maritime Law – Policy; Consideration; Comparison' (2001) 66 J. Air L. & Com. 689 at 693; G.L.H. Goo, 'Deregulation and Liberalization of Air Transport in the Pacific Rim: Are They Ready for America's 'Open Skies?' (1996) 18 U. Haw. L. Rev. 541 at 548.

[25] G. Lipman, 'Multilateral Liberalization – The Travel and Tourism Dimension' (1994) 19 Air & Space L. 152 at 153.

[26] de Palacio 2001, *supra* note 1.

[27] Lipman, *supra* note 25 at 152.

[28] *North American Free Trade Agreement Between the Government of Canada, the Government of Mexico and the Government of the United-States*, 17 December 1992, Can. T.S. 1994 No. 2 (1993) 32 I.L.M. 289 (entered into force 1 January 1994) [hereinafter *NAFTA Agreement*].

[29] The Association of Southeast Asian Nations (ASEAN) was established on 8 August 1967 by the five original Member countries, namely, Indonesia, Malaysia, Philippines, Singapore, and Thailand. Brunei Darussalam, Vietnam, Laos, Myanma, and Cambodia joined the

of these agreements to cooperate in economic fields, the completion of the European single aviation market, and the emergence of similar regional affiliations within Central and South America, the potential is there for groups of States to negotiate multilaterally – block-to-block negotiations – offering their airlines baskets of opportunities to a range of markets. However, only if national restrictions on ownership and control are lifted will airlines of different States be able to fully invest in one another.

### 2.3 *Sustainability of Air Carriers in the Current Global Market*

A complete study of the new aviation global market requires a short presentation of the main current challenges of air carriers. Thus, before studying our main subject – the question of airline ownership and control – the concept of 'sustainability of airlines', as one of the main current concerns, should be addressed.

Three major categories of air carriers can be distinguished today. The first consists of the large network carriers (i.e. Delta, American Airlines, United Airlines in the US; British Airways, Air France, Lufthansa in the EU). These are recognized mainly by their strong home market, their global reach and ability to bring passengers within their network to a large variety of destinations. They employ a mix of different aircraft types, with hub and spoke networks, and with an expensive selling and marketing structure using CRSs and travel agents. The second category comprises the middle-size carriers with a more limited network, with restricted or no home market (i.e. KLM and SAS in Europe). The third category is the low-cost carriers essentially for regional travel, providing a low-cost product, with one single aircraft type, often using secondary airports, with simple fare structure available on the internet only and with limited connecting possibilities (i.e. Ryanair, EasyJet and Buzz in Europe; South-west Airlines, Jet Blue and WestJet in North America; Virgin Blue and Australian Airlines in Australia, Air Asia in Malaysia). These three categories have unequal positions in the market. The strong large network carriers, in parallel with the fast development of low-cost carriers these past years, make the mid-sized airlines more and more fragile and unable to react adequately to new competitive challenges. Consequently, this evolution tends to place in question the sustainability of tens of air carriers worldwide that live under constant threat of financial difficulties, or even of sudden collapse, as recently demonstrated with Sabena in Belgium.

Two major reasons lie behind the increasing uncertainty of the airline industry. The first is foreign investment restrictions. The present study tends to demonstrate, *inter alia*, that national restrictions on investment weaken considerably the airline industry as a whole. As stated above,[30] airlines are increasingly buying interests in, or selling interests to, each other in accordance

---

Association afterwards. The Treaty of Cooperation in Southeast Asia was signed on 24 February 1976 and declared that in their relations with one another, the High Contracting Parties should be guided by common principles. *Inter alia*, ASEAN economic cooperation covers today many areas, including transportation, communication, and tourism.

[30] See introduction, para. 3, above.

with the general business trend. Foreign capital resources benefit national airlines, since they increase trade and jobs, help airlines to remain competitive in the worldwide market.[31] Thus, if airline cross-border investments are limited, carriers must rely on their own financial resources. Since airlines are not very profitable companies by nature, their survival is increasingly threatened. The second reason behind the serious threat to mid-sized carriers is the growing importance of low-cost carriers. Without a strong financial structure and an original service responding to a specific passenger category, mid-sized carriers cannot compete in the long-run. Indeed, it is foreseen that by around 2010–2012, only the larger carriers will remain as alternatives to the low-cost carriers which will represent about 35 per cent of the market share.[32]

Why is the disappearance of carriers such a big concern, given that we are in favor of a more concentrated competitive industry?[33] In fact, a flag carrier is a central actor in the national economy, in particular in terms of direct employment – personal of the carrier itself, and indirect employment – personal of all the activities created by the air operations (i.e. airports personal including safety, ground handling, store staff). For instance, the considerable loss of direct and indirect jobs due to the collapse of Sabena has been a real disaster in Belgium. Moreover, European and national economic markets might suffer from the decrease of route operations, since mid-sized carriers are essential to ensure an inter-continental network. Since low-cost carriers can only be profitable on thick routes, passengers living in smaller communities may well see their travel options seriously reduced, or become substantially more expensive. Indeed, neither the network carriers, nor the low-cost carriers will be ready to provide services to these destinations, or only for a very high price.

The events of September 2001 have shown the fragility of the industry and have led to emergency interventions by governments in order to avoid a total collapse of the air transport system. The entire industry is presently involved in the identification of economic, legal and technical solutions able to sustain a strong and secure airline industry. To this aim, few directions for the industry have been identified so far. Regarding the safety and security issue, the need is to build the global aviation system under the conditions of a new world instability.[34] From an

---

[31] The issue regarding the need for outside capital in national airlines is developed in Part 2, Chap. 3, para. 2, below.

[32] U., Binggeli, L., Pompeo, 'Hyped hopes for Europe's law-cost airlines' (2002) The McKinsey Quarterly 4, online The McKinsey Quarterly
http://www.mckinseyquarterly.com/article_page.asp?ar=1231&L2=23&L3=79&srid= (date accessed: 30 October 2002).

[33] The present analysis recognized the necessity of a concentrated industry.

[34] Philip A. Karber, 'Re-constructing Global Aviation in an Era of the Civil Aircraft as a Weapon of destruction' (2002) 25 Harv. J. L. & Pub. Pol'y 781, at 807. To reconstruct the global aviation system, P. Karber states that the main ideas are 'to level the playing field for all countries, in terms of access to advanced security tactics, technology, and training', 'not to redefine norms, but for more effective international law' (which implies enforcement of effective control and sanctions), and 'to reconsider the rules that constitute how hijacking is treated'.

economic point of view, new strategies and new airline models would help the business recover.[35] However, any recovery will have to come from long-term, structural cost reductions. For major airlines using the high-coverage hub-and-spoke model, such reductions may be difficult to achieve, and these airlines may struggle beyond 2004. In contrast, as stated earlier, competitors using a lower-cost strategy appear well positioned to expand their operations and profitability. The industry that recovers from these challenges is likely to look very different from today´s industry. On the one hand, despite the pressure on traditional players, the hub-and-spoke model will probably not disappear, as it will remain necessary for connectivity between smaller airports. On the other hand, while the lower-cost strategies can be ideal for leisure travelers, they are not suitable for many business fliers, because they cannot offer the flexibility, range of flights and destinations, or level of service that business travelers demand. Therefore, the business airline model seems to have a positive future as well.

This likely evolution of the aviation industry requires an effort of the national and international authorities, and more important, an effort of the air carriers themselves to adapt their services to the increasing specific demands. Low-cost or high service quality, operations on major routes or on thin regional routes are strategies that air carriers have to choose between in order to survive and succeed in the current aviation global environment. This analysis leaves the mid-sized carriers apart, although their survival is clearly essential for the national economies. Most of these carriers are willing to adapt to the specific demand to sustain on the regional and international market, however, they cannot do it without States' endeavors. Liberalization of ownership and control restrictions is, again, the solution for these carriers' uncertainty. Indeed, if they were allowed to merge with, or to be taken-over by, a large network carrier, they would get the financial means necessary to continue to operate.

### 3. Alliances, Open-Skies, Full Market Access: a Three-stage Process to Achieve Globalization

Airlines progressively build the 'full liberalization' of the industry; in other words, complete abolition of all national impediments to the consolidation and globalization of the air market. Due to the complexity of the process, airlines have attempted to implement a progressive liberalization and the elimination of regulatory constraints incrementally. Whereas the alliance phenomenon constituted the first sign that the current restrictions were outdated, Open-Skies agreements are the first step in the process of full liberalization.

---

[35] Peter Costa, Doug Harned, Jerrold Lundquist, 'Rethinking the aviation industry' (2002) The McKinsey Quarterly 2, online The McKinsey Quarterly
http://www.mckinseyquarterly.com/article_page.asp?ar=1190&L2=23&L3=79 (date accessed: 13 May 2002).

*3.1 Alliances: a Strategic Means to Avoid Foreign Investment and Traffic Rights Restrictions*

'Alliance' is not a legal term in the context of aviation, and because of the existence of all kind of different cooperative arrangements in the airline industry, it is hard to define. In general terms, an alliance is a commercial agreement, where reciprocal rights are negotiated, between two or more airlines from the same State or different States, though the purpose of the agreement can vary from one arrangement to the next.[36] Some have tried to distinguish between 'market oriented alliances' that aim at increasing traffic and market share (e.g., code-share agreements, hub coordination, block spacing, and FFP agreements) and 'cost oriented alliances' that aim at reducing cost (e.g., joint ventures, reciprocal sales, catering and maintenance, and asset sharing).[37] Other distinctions between types of alliances have also been made.[38] In any case, the advent of the 'global alliance', caused by, *inter alia*, the national restrictions on ownership and control of airlines, is considered one of the major commercial developments of the last decade[39] and will, therefore, merit special attention.

Foreign investment and traffic rights restrictions have long been predominant issues in the air transport sector; however, there are two main reasons for the emergence of airline alliances since the mid-1990s.[40] First, forming alliances allowed airlines to expand their markets and receive additional benefits of enhanced presence and customer loyalty, while reducing their capital expenditures and their overall costs through the more efficient use of assets, especially aircraft.[41]

---

[36] For an overview of airline alliances, see J. Naveau, 'Les Alliances entre Compagnies Aériennes. Aspects Juridiques et Conséquences sur l'Organisation du Secteur' (1999) 49 ITA Etudes & Doc. 9; for an analysis of the different degrees of commitment of airline alliances, see J. Balfour, 'Airline Mergers and Marketing Alliances – Legal Constraints' (1995) 20 Air & Space L. 112 at 112.

[37] R. Polley, 'Defense Strategies of National Carriers' (2000) 23 Fordham Int'l L.J. 170 at 195.

[38] For the distinction between 'strategic alliances' and 'tactical alliances', see *ibid.*; for the distinction between 'equity alliances' and 'joint venture alliances', see M.S. Simon, 'Aviation Alliances: Implications for the Quantas – BA Alliance in the Asia Pacific Region' (1997) 62 J. Air L. & Com. 841 at 843; for the description of 4 types of alliances (interlining, joint operations, code sharing, franchising), see Balfour, *supra* note 36 at 116.

[39] For an overview of airline alliances worldwide, see 'The Global Alliance Grouping' *Airline Bus.* (May 2000) 59; about Star Alliance, see N. Ionides, 'Expanded Horizons' *Airline Bus.* (Nov. 1999) 34.

[40] According to WTO, 'ICAO attributes the recent development of alliances to factors such as: globalization of business practices and attitudes; existing market access constraints resulting from the bilateral system; liberalization trends at domestic and regional levels; and commercial incentive related to economies of scope and scale', see WTO doc.1, *supra* note 8 at 9.

[41] R.I.R. Abeyratne, *Emergent Commercial Trends and Aviation Safety* (Aldershot: Ashgate, 1999) xiii [hereinafter Abeyratne *Emergent Commercial Trends*]; about 'The Philosophy of Strategic Alliances', see Abeyratne, *Aviation Trends in the New Millennium* (Aldershot: Ashgate, 2001) [hereinafter Abeyratne *Aviation Trends*]; M.J. AuBuchon, 'Testing the

Second, and most importantly, forming alliances allowed international airlines to circumvent national restrictions on ownership and control and make a more efficient use of existing traffic rights. Indeed, in the face of increased international competition and ownership restrictions that precluded mergers, alliances were the sole means by which airlines could consolidate market share. As Abeyratne so aptly stated: '[t]he reason for airlines banding together is to share an otherwise wasted market which is still regulated by bilateral governmental negotiations.'[42]

Some of today's alliances go far beyond mere cooperation, as they go a long way down the road towards full integration.[43] Not surprisingly, the alliance has for the moment taken the place of the airline merger, since it affords airlines the means to avoid ownership restrictions central to the bilateral system, but at the same time, enjoy the benefits of market consolidation. Alliance relationship is sometimes anchored by a co-ownership of assets or through mutual equity investments.[44] Thus, alliances have become more and more strategic, they have given rise to competition law concerns,[45] mainly due to the conflicts over various applications of divergent competition laws to airline alliances, essentially the divergences between EC competition law and US antitrust law.[46] In any case, though the global airline alliance has proven profitable for more than ten years, it is not a long-term solution for economic integration of the industry.[47] This can only be achieved through international merger, which will only be possible when national authorities eliminate restrictions on foreign investment in domestic airlines.

---

Limits of Federal Tolerance: Strategic Alliances in the Airline Industry' (1999) 26 Transp. L.J. 219 at 220; S. Mosin, 'Riding the Merger Wave: Strategic Alliances in the Airline Industry' (2000) 27 Transp. L.J. 271 at 272; C. Tarry, 'Playing for Profit' Airline Bus. (June 1999) 90.

[42] Abeyratne Emergent Commercial Trends, ibid. at xii; the main current example is the British Airways/American Airlines alliance proposal; the proposal still has to be approved by the US Department of Justice (DOJ), see e.g. Department of Justice, Immediate Release, 'Justice Department Urges DOT to Impose Conditions on American Airlines/British Airways Alliance' (17 December 2001), online: DOJ
http://www.usdoj.gov/atr/public/press_releases/2001/9705.htm (date accessed: 18 December 2001).

[43] 'KLM and Northwest, who have now been partners for the best part of a decade, have clearly gone further than in integrating operations across the Atlantic', see C. Baker, 'Behind The Handshake' Airline Bus. (February 2001) 66 at 68.

[44] AuBuchon, supra note 41 at 221.

[45] Polley, supra note 37 at 196.

[46] A.C., Lu, International Airline Alliances : EC Competition Law/ US Antitrust Law and International Air Transport (The Hague: Kluwer Law International, 2003).

[47] Airline alliances have been criticized by a number of authors, as being a 'poor man's merger', see e.g. IATA doc., supra note 13 at 31; Airline alliances are 'second-best solutions and do not provide the economic gains tied to tighter integration', see P. Sparaco, 'European Deregulation Still Lacks Substance', Aviation Wk & Space Tech. (9 November 1998) 53.

*3.2 Open-Skies Agreements: a Bilateral Liberalization*

The US Open-Skies initiative can be viewed as an important step forward in the process of airline liberalization, as it eliminates one of the most significant impediments to full liberalization of the airline industry by liberalizing traffic rights between the partners. Indeed, 'Open-Skies agreements allow unrestricted service by the airlines of each side to, from and beyond the other's territory, without restrictions on where carriers fly, the number of flights they operate, or the prices they charge.'[48] The US Open-Skies initiative, which was first announced in a 1992 DOT Order (dated 5 May 1992),[49] was intended to establish Open-Skies as a worldwide regime. The Open-Skies policy is defined in the DOT Order of 5 August 1992[50] by 11 provisions designed to ease restrictions on the aviation relationship between the US and any other signatory State. For instance, Open-Skies permits open entry on all routes to and from the US and the other party, unrestricted capacity and frequency on all routes, and unrestricted route and traffic rights.[51] '[T]he US government intends to build on the progress in establishing Open-Skies agreements with its aviation partners',[52] by following the

---

[48] Department of Transportation, Immediate Release 68-01, 'US Secretary of Transportation Says Bush Administration to Press for Global Aviation Liberalization' (30 June 2001) [hereinafter DOT doc. 68-01].

[49] Department of Transportation, *Order in the Matter of Defining 'Open-Skies'*, 57 Fed. Reg. 19323-01, DOT Order No. 92-4-53 (5 May 1992).

[50] Department of Transportation, *Order in the Matter of Defining 'Open Skies'*, DOT Order 92-8-13, Docket 48130 (5 August 1992) [hereinafter DOT Order 92-8-13].

[51] These three elements are the three first provisions defined by the DOT Order. The eight last elements are double disapproval pricing in Third and Fourth Freedom, liberalization of the charter arrangements and cargo regimes, ability to convert earnings and remit in hard currency promptly and without restriction, the right for the carriers to perform their own support functions, the guarantee of fair competition, and explicit commitment for non-discriminatory operation of and access to computer reservation system. For an analysis of each provision, see *ibid.* at 3; A.L. Schless, 'Open Skies: Loosening the Protectionist Grip on International Civil Aviation' (1994) 8 Emory Int'l L. Rev.435 at 447.

[52] The last Open-Skies agreement negotiated by the US was signed by Bush Administration on 19 October 2001 between the US and France, see Department of Transportation, Immediate Release 111-01, 'United-States, France Reach Open-Skies Aviation Agreement' (19 October 2001); On 16 June 2001, between the US and Poland, informal discussions were held with the UK and with Hong-Kong, negotiations have started with Japan, and an expanded policy is foreseen with African countries (in 2000, agreements were signed with Senegal, Benin, Rwanda, Morocco, and Nigeria), see DOT Doc. 68-01, *supra* note 48; an agreement was signed between the US and Portugal last year, see F. Fiorino, 'More Open Skies', *Aviation Wk & Space Tech.* (12 June 2000) 19; about the US Open-Skies policy in Asia, see Goo, see *supra* note 24 at 542; and the US still tries to replace the UK–US bilateral air services agreement with an Open-Skies agreement, see Abeyratne *Emergent Commercial Trends, supra* note 41 at xiv.

KLM/Northwest arrangement, in which DOT approval of the airline alliances was conditioned upon the Netherlands' acceptance of an Open-Skies agreement.[53]

Although US implementation of its Open-Skies policy has been very successful, with many countries having already agreed to this regime,[54] it does not represent real progress towards full liberalization of the international air transport industry. While the US Secretary of Transportation asserts that 'the Bush administration is committed to eliminate barriers to free trade in aviation services across the globe,'[55] Open-Skies will only liberalize market access with their partners and neither foreign investment nor cabotage is addressed under the US Open-Skies agreements.[56] As de Palacio stated in her speech on globalization: '[t]hese Open-Skies agreements do not really pave the way for responsible globalization, but continue to be bilateral in nature.'[57] Thus, as she explained, the US Open-Skies policy cannot lead to full liberalization since the traditional ownership and control requirement is maintained and designation of foreign owned airlines is prohibited.

While the airline industry is in urgent need of change, the liberalization process is, at best, inching forward. 'Open-Skies', though setting a liberal standard, is not the answer as market access is only partially liberalized. Indeed, any flight from an Open-Skies signatory State to a non-signatory State is subject to the terms of the traditional bilateral agreement between the third State and the carrier's national State, and the requirement of national ownership and control is still fully in place. Consequently, Open-Skies is a step in the right direction but not enough because of its bilateral nature and because of inclusion of traditional ownership and control provision.

### 3.3 Achieving a Complete Liberalization by the Advent of Global Market Access

Clearly, the goal of the airline industry must be complete multilateral liberalization if international air carriers are to consolidate their markets and remain competitive. The 1994 ICAO Conference on International Air Transport Regulation identified a

---

[53] Goo states that 'The recent alliance of Northwest Airlines and the Netherlands' KLM is an excellent example, with both carriers able to fly without restriction into the other's markets, as provided in the 1992 US–Netherlands Open-Skies agreement, and with both also sharing the benefits of the alliance in jointly setting prices and market strategy', see Goo, *supra* note 24 at 563; and see H.A. Wassenbergh, 'Future Regulation to Allow Multi-National Arrangements Between Air Carriers (Cross-Border Alliances), Putting an End to Air Carrier Nationalism' (1995) 20 Air & Space L. 164 at 164.

[54] Between 1995 and December 2000, nearly 80 Open-Skies bilateral agreements had been concluded (36 in the last 3 years) between approximately 60 countries, see ICAO, *Annual Report of the Council*, ICAO Doc. 9770 (2001) 5 [hereinafter ICAO Annual Report 2001].

[55] DOT Doc. 68-01, *supra* note 48.

[56] A. Edwards, 'Foreign Investment in the U.S. Airline Industry: Friend or Foe?' (1995) 9 Emory Int'l L.R. 595 at 609; WTO, *Note on Developments in the Air Transport sector Since the Conclusion of the Uruguay Round, Part Four*. WTO Doc. S/C/W/163/Add.3 (2001) 21 [hereinafter WTO doc.2].

[57] de Palacio 2001, *supra* note 1.

number of measures that could be undertaken to achieve unrestricted market access and, ultimately, complete liberalization of the international air transport industry:

> Entering into an agreement or agreements which liberalize(s) blocks of market access (such as the all-cargo market or the non-scheduled market, prior to consideration of one for scheduled passenger operations);
> Providing a macro-level guaranteed periodic incremental increases not tied to market growth;
> Reducing or eliminating over time existing impediments to inward (foreign) investment in national air carriers and having the right of establishment of air carriers;
> Initially fully liberalizing basic market access for services touching the territories of both the granting and receiving parties, then optionally phasing in so-called Seventh Freedom and/or cabotage rights at future times.[58]

Accordingly, full market access can be realized by removing all the national restrictions, which means liberalization of traffic rights among States, not just between parties to bilateral agreements (even the US Open-Skies agreements), and abolishment of the principle of substantial ownership and effective control of airlines. This concept of full market access neither contradicts the Open-Skies policy nor undermines it; rather it goes further, establishing a fully integrated open market in which all air carriers would be free to operate and provide services in all participating countries. Implementation of the ICAO framework has begun, but only on a regional level – for example, the EU internal market has been fully liberalized between the Member States since 1997.[59] These recommendations will not be implemented on a global scale until national authorities recognize that complete liberalization is the long-term prerequisite for the growth of the airline industry.

## 4. Conclusion

This chapter can, nevertheless, conclude on a positive note. The airline industry is an industry full of paradoxes: restrictions on foreign investments in an increasingly global air market, the contradiction between nationality of airlines and their international activities, and airlines operating on a multilateral and global scale governed by bilateral agreements that only look at the market between two States.[60] However, signs of progress can be seen among the aviation players since the last

---

[58] ICAO, Working Paper (*World-Wide Air Transport Conference on International Air Transport Regulation: Present & Future*) No. AT Conf/4 – WP 7 (18 April 1994) 8 [hereinafter ICAO Working Paper No. ATConf/4 – WP 7].

[59] About the EU Liberalization, see *supra* note 4; for further analysis on the EU liberalization see Part 1, Chap. 2, para. 3, below.

[60] This last paradox has been stressed by Ms de Palacio last May at the conference on air transport Globalization, see L. de Palacio 2001, *supra* note 1.

ICAO Air Transport Conference of 1994.[61] Indeed, at that time, most governments were not prepared to abandon national ownership restrictions, due primarily to concerns over the risks of weakening their sovereignty, as, for States, having an airline owned and controlled by national interests was a sign of prestige and independence. Today, fewer States continue to shy away from the liberalization of international air transport. The process of achieving globalization is underway and the air transport landscape now stands 'somewhere' between one of three stages: (1) the alliance phenomenon of the early 1990s, which is still ongoing; (2) the US Open-Skies policy which began in 1992, and has yet to be implemented in some major aviation markets (e.g., the English market); and (3) full market access, which to date has only been implemented on a regional basis.

---

[61] C. Thornton & C. Lyle, 'Freedom's Paths' *Airline Bus.* (March 2000) 74.

# Chapter 2

# The Progressive Decline of National Regimes on Ownership and Control of Airlines

## 1. Introduction

Transnational investment is a fact that lawmakers cannot ignore. According to the ICAO Air Transport Bureau, foreign investors, including foreign air carriers, owned 166 of the 984 air carriers operating worldwide in 2001. This trend affects the airlines of developing countries and developed countries alike,[1] since foreign ownership of all airlines has increased.[2] Yet despite this apparent evolution, States have generally maintained foreign ownership limitations on airlines,[3] and while there are progressively more signs of relaxation of these restrictions, it is unlikely that the 'ownership and control principle' will be abandoned anytime soon. Thus, the paradox of the airline industry remains firmly entrenched, particularly in the US where national restrictions on airline ownership are still very strict.

## 2. The US Exception: Perpetuating Protectionism Despite DOT's Willingness to Liberalize

America 'should not be an orphan' in the internationalization of the aviation industry. However, while the US air transport regulators (i.e., the DOT) try to

---

[1] WTO, *Note on Developments in the Air Transport Sector Since the Conclusion of the Uruguay Round, Part Five*. WTO Doc. S/C/W/163/Add.4 (2001) 5 [hereinafter WTO doc.1].

[2] *Ibid.* at 15; J.D. Brown, 'Foreign Investment in U.S. Airlines: What Limits should be Placed on Foreign Ownership of U.S. Carriers?' (1990) 41 Syracuse L.R. 1269 at 1270.

[3] For an overview of the national restrictions all over the world, see WTO Doc.1, *ibid.* at 20 (ICAO and IATA source, 1 October 2000); see also IATA, Government and Industry Affairs Department, *Report of the Ownership & Control Think Tank World Aviation Regulatory Monitor*, IATA doc. prepared by H.P. van Fenema (7 September 2000) at 43 [hereinafter IATA doc.] (result of the survey led by IATA in 2000, a questionnaire was used to compile a country-by-country overview for about 30 States, plus the European Economic Area).

adapt US policy to meet global economic needs, US law continues to limit foreign ownership in the nation's airline industry.[4]

## 2.1  The Protectionist US Law

The law on foreign ownership has been a major concern for the US Congress since the commercial aviation industry first took shape in the late 1920s. It is interesting to analyze briefly the three stages of the ownership law, as it has been continuously strengthened since that time. The law was born out of the US national security concerns that predominated the late 1920s and thereafter evolved to meet US economic and political ends. Initially, national ownership restrictions were imposed on many US industries that were deemed essential to national security. Since the aviation industry was directly involved in national defense,[5] the 1926 *Air Commerce Act*[6] was enacted, in part, to place limitations on foreign ownership of US air carriers. Notably, this Act was the first regulation of the airline industry as a whole, and was thus also intended to foster development of the fledgling industry.[7] The 1926 Act stated that aircraft could be registered in the US only if owned by US citizens; it further required that US citizens control at least 51 per cent of the voting interest of any US air carrier, and that the carrier's president and at least two-thirds of its board of directors be US citizens.[8] After the Great Depression of the 1930s, the state of the economy took its place as a major element of US national security and law-makers chose 'protectionism' as the primary means of safeguarding the nation's airline industry.[9] The *Civil Aeronautics Act* of 1938[10]

---

[4] The US has adopted a very restrictive regulation on foreign ownership in different sectors, see Abeyratne, *Aviation Trends in the New Millennium* (Aldershot: Ashgate, 2001) at 362-364 [hereinafter Abeyratne *Aviation Trends*].

[5] For further information about national defense concerns of the US in the 1920s, see S.M. Warner, 'Liberalize Open Skies : Foreign Investment and Cabotage Restrictions Keep Non Citizens in Second Class' (1993) 43 Am. U. L. Rev. 277 at 305; K. Bohmann, 'The Ownership and Control Requirement in U.S. and European Union Air Law and U.S. Maritime Law – Policy; Consideration; Comparison' (2001) 66 J. Air L. & Com. 689 at 696.

[6] *US Air Commerce Act*, Pub. L. No. 69-254, SS 1-14, 44 Stat. 568 (1926).

[7] D.T. Arlington, 'Liberalization of Restrictions on Foreign Ownership in U.S. Air Carriers: the United States must take the First Step in Aviation Globalization' (1993) 59 J. Air L. & Com. 133 at 141.

[8] The *Air Commerce Act*, ch. 344, § 3(a), 44 Stat. 568, 569, and § 9(a), 44 Stat., *supra* note 6. The Act defines a US citizen as: (1) an individual who is a citizen of the United States or its possession, or (2) a partnership of which each member is an individual who is a citizen of the United States or its possessions, or (3) a corporation or association ... of any State, Territory, or possession thereof, of which the president and two-thirds or more of the board of directors or other managing officers thereof, as the case may be, are individuals who are citizens of the United States or its possessions and in which at least 51 per cent of the voting interest is controlled by persons who are citizens of the United States or its possessions.

[9] For further information about the US economic protectionism since the 1930s, see Bohmann, *supra* note 5 at 696; see A. Edwards, 'Foreign Investment in the U.S. Airline Industry: Friend or Foe?' (1995) 9 Emory Int'l L.R. 595 at 603; see C.G. Alexandrakis,

thus modified the *Air Commerce Act* by strengthening restrictions on foreign ownership of US airlines: the statute increased the minimum percentage of US citizen-held voting equity required for US air carriers from 51 to 75 per cent,[11] and left the US citizenship requirement intact.

Today, the *Federal Aviation Act* of 1958 governs the US airline ownership regime.[12] Enacted during the tense climate of the Cold War, the Act further narrowed citizenship restrictions for owners of US air carriers and thereby restricts who may operate commercial aircraft in the United States. It requires that anyone wishing to operate aircraft within the US must first apply for and obtain a 'certificate of public convenience' from the DOT and further provides that this certificate can only be issued to an air carrier that is a 'citizen of the United States.' Section 1301(16) of the Act defines a 'US citizen' as:

> (a) an individual who is a citizen of the U.S. or one of its possession, or
> (b) a partnership of each member is such an individual, or
> (c) a corporation or association created or organized under the laws of the U.S. or of any State, Territory, or possession of the U.S., of which the president and two-thirds or more of the board of directors and other managing officers thereof are such individuals and in which at least 75% of the voting interest is owned or controlled by persons who are citizens of the U.S. or one of its possessions.[13]

The US *Federal Aviation Act* has remained essentially unchanged since 1958, 'at least as far as the written law is concerned.'[14] However, the law has been criticized as ambiguous in a number of key respects. First, the US citizenship provision lacks requisite specificity. For example, the statutory definition of citizenship refers to partnerships, but does not address the question of whether 'a partnership' includes only individual persons or it includes corporate partners as well?[15] Second, the law contains no objective standard for what constitutes 'effective control' of an airline; hence, US regulators are free to subjectively interpret this notion according to prevailing US interests. Indeed, neither the CAB nor DOT has ever established a clear definition of 'effective control'.[16] The issue of control as it relates to the percentage of non-voting shares of an airline that may be held by foreigners thus

---

'Foreign Investment in U.S. Airlines: Restrictive Law is Ripe for Change' (1994) 4 U. Bus. Miami L.J. 71 at 74.

[10] *Civil Aeronautics Act, supra* note 6.

[11] *Ibid.*, ch. 601, § 1(13), 52 Stat. at 978.

[12] Followed by the *Airline Deregulation Act*, Pub. L; No. 95-904, § 102(7), (10), 92 Stat. 1705 (codified as amended at 49 USC § 1301-1552 (1982)) [hereinafter the *Airline Deregulation Act*].

[13] *Ibid.*, 49 U.S.C. app.§ 1301(16) (1988).

[14] Arlington, *supra* note 7 at 142.

[15] About the uncertainty of the word 'partnership' and the whole interpretation of the citizenship requirement, see J.T. Stewart, 'U.S. Citizenship Requirements of the Federal Aviation Act – A Misty Moor of Legalisms or the Rampart of Protectionism' (1990) 55 J. Air L. & Com. 685.

[16] Warner, see *supra* note 5 at 307.

remains 'a matter of policy, not law.'[17] Consequently, US restrictions on 'ownership and control' of airlines have actually been tightened through the broad discretionary powers that the statute affords US regulatory authorities.[18]

---

[17] H. Wassenbergh, 'Towards Global Economic Regulation of International Air Transportation through Inter-Regional Bilateralism' The Hague (August 2001) at 7 [Unpublished].

[18] As regard the DHL case, the DHL Airways's ownership and control remains very questionable, despite the DOT's decisions in favor of DHL. In January 2001, in light of the strong possibility that DHL Airways may be under the control of foreign nationals, Federal Express Corporation (FedEx) requested the DOT to conduct a formal investigation into the compliance of DHL Airways with the statutory citizenship requirements applicable to all US air carriers, see Department of Transportation, Order dismissing *Third-Party Complaint of Federal Express Corporation in Docket OST-t-2001-8736 and of United Parcel Service Co. (in Docket OST-2001-8824) without prejudice Grant the Motions to File Otherwise Unauthorized Documents Filed by Federal Express Corporation and DHL Airways Inc.*, DOT Order 2001-5-11, Docket OST-01-8736-8 (11 May 2001), online: DOT http://152.119.239.10/docimages/pdf58/120921 web.pdf (date accessed: 14 May 2001), the department dismissed the complaint over DHL Airways' citizenship (more precisely, it dismissed FedEx's petition to revoke DHL Airways' authority to operate scheduled all-cargo service between the United States and Kuwait). In its decision, the DOT emphasized: 'The regulations permitting foreign air freight forwarders to operate in the US were created to 'eliminate the citizenship barrier to entry, promote competition among indirect air carriers, increase business on US air carriers, and reaffirm the US commitment to promote competition in the air transportation industry.' The Department's Decision affirming the DHL WE license accomplishes these objectives, see Department of Transportation, Immediate Release 45-01, 'DOT Rules on Petitions Against DHL' (11 May 2001), online: DOT http://www.dot.gov/affairs/dot45-01.htm (date accessed: 1 October 2001); DHL WE, Immediate Release, 'DHL Worldwide Express Welcomes DOT Ruling' (11 May 2001), online: DHL WE http://www.dhl-usa.com/press_display/1,3574,79,00.html (date accessed: 1 October 2001); DOT's related decisions are the following: Department of Transportation, *Application of DHL Airways, Inc. pursuant to 49 U.S.C. Section 40109(c) – Exemption – U.S.-Kuwait via Brussels and Bahrain*, Docket No. OST-2000-6937 (14 February 2000); Department of Transportation, *Application of the Registration of DHL Worldwide Express, Inc., as a Foreign Air Freight Forwarder*, Docket No. OST-2000-8732-1 (10 October 2000); Department of Transportation, *Application of Federal Express Corporation against DHL Airways, Inc. Regarding Compliance with U.S. Citizenship*, Docket No. OST-2001-8736 (19 January 2001); these three last decisions can be found online: DOT http://dms.dot.gov/search/hitlist.asp (date accessed: 4 October 2001). In 2002, DHL Airways has been subject to further investigation since the two main competing carriers (FedEx, one more time, and United Parcel Service company (UPS)) filed two separate Petitions for Reconsideration, with the DOT, to institute a public inquiry into the citizenship and foreign control of DHL Airways. The DOT issued in that respect a notice consolidating these two petitions in Docket OST-2002-13089. In fact, in May 2002, after its first investigation, the DOT found that DHL Airways was actually controlled by US citizens and met all statutory tests, whereas the DHL's ownership remains highly doubtful. Indeed, in July 2002, the Deutsche Post announced its intention to increase considerably its shares in the US carrier (see 'DHL International – 100% du capital pour la Deutsche Post' L'Antenne (6 December 2002)). The question to be raised is whether DHL Airways passed the DOT control test

## 2.2 DOT Discretion - Overstepping the Bounds of the Federal Aviation Act

Since 1958, the CAB and then the DOT[19] have employed a 'two-pronged' approach to the *Federal Aviation Act*'s citizenship requirement: first, to qualify as a 'US citizen,' an airline must satisfy the Act's US ownership percentages (§ 1301(16)(c)); and second, only the airlines that can qualify as a 'US citizen' may 'control' a US air carrier. The latter condition is particularly vague, as the law does not define what constitutes 'control' and, thus, the notion of 'control' has been susceptible to varying interpretations, based upon the policy goals of the administration in place at the time.

*2.2.1 Adapting air transport policy to meet US economic and political ends* The US made its ownership and control policy evolve according to the international and national airpolitical situation. Thus, the US policy can be divided into two phases: the pre-1989 period and the post-1989 period.

The first period was dominated by three decades of Cold War. In the 1960s, DOT policy was dominated by political suspicion. The US took a guarded approach *vis-à-vis* those countries it viewed as susceptible to communist influence, lest air traffic rights be granted to a country that could suddenly move into the enemy camp. Consequently, the DOT narrowly interpreted citizenship restrictions applicable to airline ownership. As in the 1920s and 1930s, national security concerns pushed the US towards protectionism, and its protectionist stance was only toughened by the good economic climate within the US airline industry. Indeed, at that time, carriers in America's burgeoning civil aviation industry were reaping the benefits of healthy competition in the domestic US market, which made them a great deal more efficient than their international competitors. Not surprisingly, DOT sought to preserve the US civil aviation industry's success, together with the abundance of capital it engendered, with a stringent interpretation of the US citizenship requirement.

Beginning in the 1960s and continuing through the late 1980s, the CAB and then the DOT applied an 'actual control' test, whereby an airline that satisfied the ownership percentage requirements of the *Federal Aviation Act* might still not qualify for US citizenship. The first case to apply the test was *Willye Peter Daetwyler, D.B.A. Interamerican Airfreight Co., Foreign Permit* (1971).[20] In addressing the issue of whether Interamerican qualified as 'a citizen of the United-States', the CAB held that the enterprise did not qualify as a US licensed air carrier: while Interamerican met the legal criteria for citizenship, it failed to

---

because of its political influence. The DOT should better further investigate such a doubtful case, if it wants to convince the entire airline industry of its clear and objective policy.

[19] About the functions and mandate of the DOT, see T.D. Grant, 'Foreign Takeovers of United States Airlines: Free Trade Process, Problems and Progress' (1994) 31 Harv. J. Legis. 63 at 69.

[20] Civil Aeronautics Board, *Order in the Matter of Willye Peter Daetwyler, D.B.A. Interamerican Airfreight Co., for Amendment of its Foreign Permit Pursuant to Section 402(f) of the FAA of 1958,* Docket No. 118, 120-21 (1971).

conform to the spirit of the statute.[21] The next major case to address the issue of 'actual control' was *Première Airlines, Fitness Investigation* (1982).[22] In *Première Airlines*, the CAB likewise focused on whether the airline was a US citizen as defined in the 1959 Act with respect to the issue of control. Once more, the CAB maintained its strict stance, demanding that Première reorganizes to address the Board's concerns over 'actual control'.[23]

US regulators took even more stringent stances in two later cases. The first was the case of *Page Avjet Corporation* (1983),[24] where the CAB held that even though all voting stock and over 75 per cent of the non-voting stock lay in the hands of US citizens, the corporation was nevertheless subject to foreign control.[25] The second of these decisions came in *Intera Arctic Services* (1987),[26] in which the DOT made clear that merely fulfilling the letter of the control/ownership statute would not render a certification applicant immune from scrutiny.[27]

After decades of strictly interpreting ownership and control requirements, the DOT finally began to temper its interpretation beginning around 1989, in response to the changing needs of the US airline industry. The DOT's change of heart reveals the political nature of the ownership and control issue and clearly demonstrates why States are so reluctant to abolish these restrictions since they can be an effective tool for furthering a State's national economic interests.

Since 1989, the world has recovered its stability. In the late 1980s, the national security threat that dominated US aviation policy throughout the Cold War began to subside and the few political rivals that remained *vis-à-vis* the United States were not strong competitors in air transport. However, the suspicion of foreigners that was initially born of national security concerns was seemingly

---

[21] The CAB held that 'where an applicant has arranged its affairs so as to meet the bare minimum requirements set forth in the Act, it is the Board's view that the transaction must be closely scrutinized and that the applicant bears the burden of establishing that the substance of the transaction is such as to be in accordance with the policy, as well as the literal terms of the specific statutory requirements,' see *ibid.* at 121; for further information about the decision, see Arlington, *supra* note 7 at 144.

[22] Civil Aeronautics Board, *Order in the Matter of Première Airlines and Fitness Investigation*, CAB Order 82-5-11 (5 May 1982).

[23] For details about the different steps of the *Première* decision, see Arlington, *supra* note 7 at 145.

[24] Department of Transportation, *Order in the Matter of Page Avjet Corporation*, DOT Order 83-7-5, Docket No. 40,905 (1 July 1983).

[25] The CAB stated, '[w]e have recognized that a dominating influence may be exercised in ways other than through a vote', *ibid.* at 3; 'The nonvoting foreign shareholders held the power to veto major company decisions, including any decisions pertaining to company consolidation, merger, acquisition, or liquidation', *ibid.* at 4; for further information about *Page Avjet Corporation* case, see Arlington, *supra* note 7 at 147, and Brown, *supra* note 2 at 1277.

[26] Department of Transportation, *Order in the Matter of Intera Arctic Services, Inc.*, DOT Order 87-8-43, Docket No. 44,723 (18 August 1987).

[27] 'If persons other than US citizens, individually or collectively, can significantly influence the affairs of [the carrier], it is not a US carrier', see *ibid.;* Arlington, *supra* note 7 at 150; Brown, *supra* note 2 at 1276.

transformed into economic suspicion and led to a fierce fare war that ultimately contributed to the deterioration of the financial health of the entire airline industry.[28] The many airline bankruptcies that followed the US airline deregulation revealed the poor state of the US civil aviation industry during this period.

In the wake of deregulation, an intense competition erupted between the 'big three' US carriers (United, American, and Delta), a few weaker US airlines, and their foreign airline competitors (mainly the European carriers). With many US airlines beset by chronic losses and urgently in need of an infusion of capital,[29] the DOT was forced to look beyond the limitations of the *Federal Aviation Act* to aid an ailing US airline industry. Protecting the nation from its political enemies was no longer a major concern of the DOT; instead, DOT's focus shifted to the formation of new economic partnerships to sustain US dominance of international civil aviation. To this end, DOT's decisions were increasingly influenced by the aviation relationship that the US was pursuing with the home country of the foreign airlines wanting to invest in US carriers. Three cases clearly illustrate this shift in priorities: the Northwest/KLM case, the USAir/British Airways (BA) case, and the Continental/Air Canada case:

• The Northwest/KLM case.
In 1989, KLM sought to make a major investment in Northwest Airlines, America's fourth largest carrier. The DOT studied this unprecedented transaction for more than three years and, in that time, altered its interpretation of the facts and reversed its own position based upon new-found economic interests. In fact, the DOT had initially refused KLM's proposal for investment in Northwest, based primarily on the *Federal Aviation Act*'s citizenship test.[30] In 1991, however, Northwest again applied to the DOT, requesting relaxation of its 1989 consent Order for the same Northwest/KLM transaction,[31] and in 1992 the DOT surprisingly granted Northwest's request.[32] In approving the transaction, the DOT considerably relaxed its own control requirements, as evidenced by the consequent protestations of other US air carriers.[33] Subsequently, the United States and the Netherlands concluded the first Open-Skies agreement, whereupon the DOT approved the Northwest/KLM request to join an alliance, i.e. to merge functions and to act as one airline, cooperating in crucial areas such as pricing, marketing and strategy.

---

[28] According to Kass, '[u]ncertainty and change have harmed the United-States airline industry in recent years', see H.E. Kass, 'Cabotage and Control: Bringing U.S. Aviation Policy into the Jet Age' (1994) 26 Case W. Res. J. Int'l L. 143 at 143.

[29] About the need for capital of the airline industry, see Grant, *supra* note 19 at 71.

[30] DOT Order 89-9-29, *supra* note 18; for more details about this decision, see Brown, *supra* note 21 at 1278; Grant, *supra* note 19 at 99; Arlington, *supra* note 7 at 152.

[31] Department of Transportation, *Order in the Matter of the Acquisition of Northwest Airlines by Wings Holdings, Inc.*, DOT Order 91-1-41 (14 January 1991).

[32] *Ibid.* at 5.; for more details about the second decision of the US DOT, see S. Mosin, 'Riding the Merger Wave: Strategic Alliances in the Airline Industry' (2000) 27 Transp. L.J. 271 at 278; Arlington, *supra* note 7 at 156; Bohmann, *supra* note 5 at 701.

[33] Grant, *supra* note 19 at 100.

• The USAir/British Airways case.

Likewise, in July 1992, British Airways (BA) announced a plan to invest in USAir, the sixth largest US air carrier, and to merge into a single brand.[34] The first BA proposal was rejected by the DOT, not only because the proposal would have given BA effective control in USAir, but also because the United Kingdom did not want to remove its protectionist barriers on US carriers' access to London. However, in March 1993 a second BA proposal was accepted by the DOT. The DOT reasoned that approval of the USAir/BA transaction would increase the likelihood of a US-UK Open-Skies agreement, though in the end no agreement was reached.[35] Indeed, to date there is no Open-Skies agreement between the US and the UK. Nevertheless, despite the collapse of the US-UK Open-Skies negotiations in October 2000, negotiations are still pending and, according to Mr. A. Sentence, the chief economist at BA, such an agreement could be a 'stepping stone' towards an EU-US common aviation area.[36]

• The Continental/Air Canada case.

The 1993 Continental/Air Canada agreement represents another example of the DOT's use of the citizenship requirement as a political and economic tool. In contrast to the two previous cases, here the US did not seek to conclude an Open-Skies agreement with Canada. Instead, the US priority was to keep its northern neighbor as a major US investment partner: '[b]ecause Air Canada's partner was a major US investment group and the US-Canada bilateral relationship was less offensive than the US-UK relationship, the control prong of the US citizenship test was not violated and the DOT ultimately approved Air Canada's proposal.'[37]

---

[34] About the first BA's proposal of July 1992, see Grant, *ibid.* at 114; Alexandrakis, *supra* note 9 at 84; Edwards, *supra* note 9 at 611; Arlington, *supra* note 7 at 158 & 173; L.R. Rose & B. Coleman, 'British Airways Buys Stake in USAir, Drawing Protests From Other Carriers' *The Wall St. J.* (22 January 1993) A3.

[35] Department of Transportation, *Order in the Matter of Joint Application of British Airways PLC for an Exemption Pursuant to Section 416(b) of the Federal Aviation Act of 1958; Application of USAir for a Statement of Authorization to offer Code-Share under 14 CFR Parts 207 and 212; Application of USAir for a Statement of Authorization for a Wet Lease,* DOT Order 93-3-17, Docket Nos. 48,634, 48,640 (15 March 1993); for further details about the second BA's proposal of March 1993, see Grant, *supra* note 19 at 128; Alexandrakis, *supra* note 9 at 88; Edwards, *supra* note 9 at 613.

[36] C. Baker, 'US and UK Remain Apart on Open Skies' *Airline Bus.* (December 2000) 19; The US remains adamant that an Open-Skies agreement is a prerequisite also for clearing the BA/AA alliance, but the UK government still hesitates to open its market, see J.D. Morrocco, 'Open Skies Impasse Shifts Alliance Plans' *Aviation Wk & Space Tech.* (November 9, 1998) 45; S. Mosin, see *supra* note 32 at 283.

[37] See Alexandrakis, *supra* note 9 at 89; for further information about Continental-Air Canada agreement, see Kass, *supra* note 28 at 175.

The major decisions of the US CAB and DOT since 1958 show how the nationality requirement of US airlines does not simply serve as a legal basis for establishing the nationality of an airline. In fact, the different application, by the US authorities, of the ownership criterion and the control criterion has to be stressed. Obviously, in their case analysis, the US apply strictly the clear and precise ownership cap of 25 per cent, while they appreciate the more flexible control criterion on a case-by-case basis. Indeed, the absence of definition of the 'control' notion left it to the varying interpretations of the administration. Thus, in each case the control notion was interpreted in a way that furthered some overarching economic and/or political objective. Early on, the DOT's rejection of merger proposals reflected US dominance in the international civil aviation sector. But, beginning with the Northwest/KLM decision, the DOT began to adopt a more flexible position with regard to foreign investment in US airlines, owing to the airlines' desire to find new financial resources to address their need for capital combined with the US need to safeguard good relations with traditional economic partners.[38] This very subjective behavior did not apply only to cases concerning the nationality of the US airlines, but as well to cases concerning nationality of foreign airlines operating into the US. In fact, the control requirement has been completely ignored in cases where there were US interests to be served. For example, in the case of the transaction between Iberia and Aerolineas Argentinas, though it was obvious that Iberia was taking the control of the Argentinian airline, the DOT remained silent in exchange for concessions from the Argentinian government that benefited US airlines operating in that country.[39]

Clearly, from these cases one may reasonably conclude that the US citizenship requirement for airlines is no longer a *per se* legal barrier to foreign investment since, at the end of the day, US regulators accept or refuse transactions based *not* on the letter of the 1958 Act, but instead on the prevailing political and economic priorities. Indeed, 'the true cause of the demise of [these] ... deal[s], was actually a conglomeration of many factors having little to do with the law itself.'[40]

2.2.2 *The necessary evolution of the US law* The domineering nature of the subjective analysis of the cases brings the fairness of the DOT's policy into doubt. The *Federal Aviation Act* strictly defines 'US citizen' and, although the DOT is allowed to apply the statute on a case-by-case basis, it must act within the limits of the law. However, the 'contradictions and uncertainty in DOT's stance towards foreign takeovers' render the letter of the law meaningless,[41] such that it is

---

[38] Even during the Cold War, the US tended to relax their policy regarding 'friendly and neighboring' countries that do not present a threat to national security, see Bohmann, *supra* note 5 at 707.

[39] IATA doc., *supra* note 3 at 25; Bohmann, *supra* note 5 at 708.

[40] Arlington, *supra* note 7 at 134.

[41] For comments on DOT's attitude by two Secretaries of Transportation, see Grant, *supra* note 19 at 101.

impossible to know with certainty the extent to which a foreign investor can invest in a US airline.

Nevertheless, perhaps despite the DOT's actions, the trend over the past few years has increasingly been towards liberalizing national ownership restrictions; so much so, that it has become clear that the time has come to drop national restrictions on ownership and control of airlines altogether. In fact, US regulators have been pressured to do so; *inter alia*, in 1992, the General Accounting Office (GAO) recognized the financial necessity of relaxing the statutory limits on foreign investment and control of airlines and in 1993, the National Commission report recommended amendment to *Federal Aviation Act*.[42]

Accordingly, it is highly recommended that the US adapt or replace the US *Federal Aviation Act* to address the current economic needs of the international air transport. As regards the ownership restrictions, the 25 per cent cap has to be raised up to 49 per cent. Such increase would allow the harmonization of the US law with the EU law. Although the long-term goal is to authorize 100 per cent of foreign investments in international airlines, the harmonization of the legal regime of the two leveraged parties, the EU and the US, is the first necessary step of the global liberalization of the ownership criterion. Regarding the control restrictions, an objective and clear legal definition of the control is the only solution to force the DOT to adopt a fair policy, more respectful of the law. Indeed, as long as the notion remains vague, the DOT' subjective behavior cannot be considered illegal.

### 3. The Two Faces of Europe: Internal EU Liberalization Versus Restrictions on Third Parties

The policy on substantial ownership and effective control of airlines among European Union Member States is less obscure and less protectionist than the US policy. However, while ownership restrictions have been totally removed within the European Union, the traditional ownership and control criteria prevail between European Member States and third countries, including the US. Moreover, the nationality requirement has not been eliminated throughout the whole European Union and the European Commission has continuously had to struggle against its own Member States on this issue. In that respect, on 5 November 2002, the European Court of Justice (ECJ) has delivered its awaited judgment on the Open-Skies cases, ruling in favor of the nationality clause replacement.

---

[42] United States General Accounting Office, *Airline Competition. Impact of Changing Foreign Investment and Control Limits on U.S. Airlines in Report to Congressional Requesters,* GAO Doc. GAO/RCED-93-7 (9 December 1992); the National Commission to Ensure a Strong Competitive Airline Industry, *Change, Challenge, and Competition: a Report to the President and Congress submitted on 19 August* 1993, Washington, DC: US Government Printing Office, 1993; for further information about these two reports, see Gertler, *supra* note 30 at 218.

*3.1 The Law Applicable to EU Member States*

The liberalization of the European system represented a major deviation from the principle of substantial ownership and effective control. Within the 'Three Packages' framework adopted by the EU Council in 1992,[43] the traditional national ownership and control requirement was replaced by the concept of 'Community Carrier.' The new rules included in the third package[44] were a product of the EU Council's call for new regulation for the licensing of air carriers within the European Community.[45] *Council Regulation 2407/92* applies to virtually all European commercial aviation carriers.[46] It provides that a Community carrier may receive an air carrier's license from its national aeronautical authority if it is majority-owned and effectively controlled by an EU Member State (and/or by nationals of an EU Member State)[47] and has its principal place of business in that Member State.[48] Carriers that meet these requirements enjoy Community status and can thus benefit from the advantages of Community legislation: e.g., the right of establishment throughout the Community and cabotage.[49]  In addition, the EU Regulation is less obscure than the US legislation to the extent that the notions of 'ownership' and 'control' have been clearly defined. First, instead of the term 'substantial ownership,' the EU Regulation uses the expression 'majority ownership.'[50]  Thus, an operating license may be granted if more than 50 per cent of the capital of the air carrier is held by any Member State or its citizens.[51] Second, unlike its US counterpart,[52] the *Council Regulation 2407/92* expressly defines 'effective control'.[53]

   *Council Regulation 2407/92* applies only between and among the 15 Member States plus Norway, Iceland, and Liechtenstein.[54] Likewise, the EU–

---

[43] The European Liberalization Regulation, see Part 1, Chap. 1, para. 1, above; for further information regarding the EU liberalization, see J., Basedow, 'Airline Deregulation in the European Community – its Background, its Flaws, its consequences for E.C.-U.S. Relations' (1994) 13 J.L. & Com. 247.

[44] EU, *Council Regulation 2407/92*, *supra* note 22.

[45] EU, *Council Regulation 2343/90*, Article 3(1), see Part 1, Chap. 1, para. 1, above.

[46] *Ibid.* at Article 1(2).

[47] *Ibid.* at Article 4(2).

[48] *Ibid.* at Article 4(1).

[49] An EU air carrier may fly any route within the EU, any international route within the EU, any domestic route within its own country, and any domestic route within another EU country.

[50] EU, *Council Regulation 2343/90*, Article 4(2), see Part 1, Chap. 1, para. 1, above.

[51] For comments given by the European Commission about the 'majority ownership', see Bohmann, *supra* note 5 at 720.

[52] About the similarities and differences of interpretation of the notion 'effective control' between the European Commission and the US DOT, see *ibid.* at 722.

[53] For the European definition of 'effective control', see Introduction, para. 2, above.

[54] The three latter States are subject to the above Community legislation by virtue of the European Economic Area (EEA) Agreement of 1994, see EU, Decision of the EEA Joint Committee 7/94 Amending the Protocol 47 and Certain Annexes to the EEA Agreement, [1994] O.J. L 160/1. The EEA Agreement is an association agreement and extends most

Switzerland agreement is an aviation-specific association agreement whereby Switzerland takes over the provisions of the EU internal air transport market.[55] Thus, between these parties, 'the traditional substantial ownership and effective control clause has become inoperative, and any challenges to an airline's traffic rights within the internal aviation market would have to be based on evidence that the airline concerned does not qualify as a Community air carrier.'[56]

### 3.2 The Law Applicable to Non-Member States

While the EU internal air transport market has been liberalized, through the 'Three Packages' Regulation, the same cannot be said for its external air transport policy. Indeed, the common European air carrier license system does not apply to air traffic between EU Member States and third countries. Instead, the EU external air transport policy is built of bilateral agreements negotiated by individual Member States, that still apply the nationality clause. The main reason behind the EU's dual system is that non-EU countries do not recognize the changes that have occurred inside the EU. Therefore, the third countries concerned are still entitled to 'withhold or revoke an operating permit if a carrier is not substantially owned and effectively controlled' by the contracting Member State and/or its nationals, according to the bilateral agreements. For example, if a carrier of Member State A is taken over by a carrier of Member State B, a third country could prevent the Member State B carrier from exploiting the traffic rights granted to carriers of Member State A by withholding or revoking the operating license granted to the carriers of Member State A. It is for precisely this reason that there have been so few mergers or takeovers between European carriers.[57]

The nationality clause in the bilateral agreements has been a source of controversy between Member States and the Commission for ten years. On the one side, the Member States were advocating a national – non-European – external air transport policy, being concerned by keeping their traffic rights outside the Community. So far most EU States have been cautious and even reticent to include the 'Community clause' in their bilateral agreements for two reasons:

(a) Such a clause, if introduced in a bilateral with a third country, would open up the benefits, obtained through the respective bilateral negotiations, to the airlines of all Member States without any guarantee that the latter would offer reciprocal benefits through similar changes of their own bilaterals concluded with third countries;

---

features of the EU internal market, including the single air transport market from the 15 EU States to these three additional European countries.

[55] An agreement on air transportation between the EU and Switzerland was signed on 21 June 1999 and entered into force on 1 June 2002. The agreement deepens relations between Switzerland and the EU, and facilitate market access for both sides.

[56] IATA doc., *supra* note 3 at 19.

[57] For instance, this problem raised between the Netherlands and Great Britain, see P.P.C. Haanappel, 'Airline Ownership and Control and Some Related Matters' (2001) 26-2 Air & Space L. 90 at 99 [hereinafter Haanappel 'Airline Ownership and Control'].

(b) The prospect for the third country concerned to have all Member States' airlines in a position where they could claim to be entitled, as a Community carrier, to enjoy the traffic rights exchanged with just one of their fellow-Members. This possibility made foreign countries hesitant to grant this concession in bilateral negotiations unless specific safeguards were introduced and, more importantly, unless counter-demands were met which would offset the value of this major concession.[58]

Today, very few Member States have actually introduced the 'Community clause'. One example is the bilateral agreement between Germany and Brunei,[59] that specifically refers to Article 4(5) of the EU *Council Regulation 2407/92*.[60] Indeed, the ownership clause grants Brunei the right to challenge traffic rights of a carrier designated by Germany only if the carrier is not able to demonstrate that it is substantially owned and effectively controlled by EU nationals.[61] And on the other side, the Commission was asking the Member States to respect the 'Community clause' requirement of the Council Regulation 2407/92. Indeed, willing to control the external air transport policy, the Commission has struggled since the early 1990s to replace all the bilateral agreements by a single agreement, to be concluded between the Community and the US. To that aim, the Commission has repeatedly sought to obtain from the Council a mandate to negotiate an air transport agreement on behalf of the Member States.

Facing the recalcitrance of the Member States to renounce their bilaterals, and, in particular, the nationality clause, the Commission filed a complaint in 1995 against eight Member States that have completed bilaterals with the US.[62] In its actions, the Commission alleges, *inter alia*, that by concluding those agreements, the States have infringed the two following obligations. First, the Commission alleges that the Member States have infringed the external competence of the Community since only the Community has competence to conclude such an agreement.[63] The Commission argues that Member States have failed to fulfill its obligations under Article 5 of the Treaty (now Article 10 of the Treaty), the Three Packages Regulations,[64] the *Council Regulation 2299/89* on Computerized Reservation Systems (CRSs),[65] and the *Council Regulation 95/93* on slots allocation at Community airports.[66] Accordingly, the Commission considers that,

---

[58] IATA doc., *supra* note 3 at 19.

[59] *Air Transport Agreement Between the Federal Republic of Germany and Brunei Darusalam*, German Federal Gazette (BGB1) 1994, II-3670, Article 3(4).

[60] EU, *Council Regulation 2407/92*, *supra* note 22 Article 4(5).

[61] For the text of the clause that Germany includes into its new or existing bilateral agreements, see IATA doc., *supra* note 3 at 20.

[62] Seven States (Belgium, Denmark, Germany, Luxembourg, Austria, Finland and Sweden) have signed 'Open-Skies' type agreements and the UK has signed a more restrictive bilateral, called Bermuda II.

[63] This complaint has not been raised against the UK/US bilateral agreement.

[64] The Three Packages Regulation, see Part 1, Chap. 1, para. 1, above.

[65] EU, *Council Regulation 3089/93 Amending Regulation 2299/89 on a Code of Conduct for Computerized Reservation Systems*, [1993] O.J. L.278/1.

[66] EU, *Council Regulation 95/93 on Common Rules for the Allocation of Slots at Community Airports*, [1993] O.J. L.14/1.

since Community air transport legislation had established a comprehensive system of rules designed to establish an internal market in that sector, Member States no longer had the competence to conclude bilateral agreements.[67] Second, the Commission alleges that the Member States have infringed Article 52 of the Treaty concerning the right of establishment by permitting the US to refuse traffic rights in its airspace to air carriers designated by the Member States which is party to the agreement, if a substantial part of the ownership and effective control of that carrier are not vested in that Member states or in its nationals.

On 31 January 2002, the Advocate General, Antonio Tizzano, has delivered his opinion on the cases.[68] Finally on 5 November 2002, the ECJ delivered its awaited judgments on the compatibility of the Open-Skies agreements with EC law.[69] It appears that the Court has upheld the Advocate General's opinion in its entirety.

Regarding the issue of Community competence, the ECJ clearly applies the *AETR* aviation doctrine that recognizes exclusive external competence for the Community in consequence of the adoption of internal measures relating to the treatment of nationals of non-member countries. Accordingly, the Court rejects the Commission's argument that it had exclusive competence to negotiate bilateral agreements as they do not affect, as a whole, the common rules, covered by *Council Regulations 2407/92* and *2408/92*, in the aviation sector. However, the Court identifies two areas of Community competence, CRSs and intra-Community fares and rates.[70]

Regarding the ownership and control issue, the Court holds that the nationality clause in the bilaterals infringes the right of establishment enshrined in Article 52 of the Treaty. Over the eight cases, the Court provides the following argumentation. Three main points can be stressed.

---

[67] Case 22/70 *Commission v Council* [1971] ECR 263 (so called the *AETR* judgment).

[68] The role of the Advocate General is to assist the Court of Justice by delivering an opinion, in other words a proposed draft decision. This opinion is not binding on the Court, however, in most cases, the subsequent Court ruling agrees with this guidance.

[69] *ECJ, Commission of the European Communities v. United Kingdom of Great Britain and North Ireland* (C-466/98), *Commission of the European Communities v. Kingdom of Denmark* (C-467/98), *Commission of the European Communities v. Kingdom of Sweden* (C-468/98), *Commission of the European Communities v. Republic of Finland* (C-469/98), *Commission of the European Communities v. Kingdom of Belgium* (C-471/98), *Commission of the European Communities v. Grand Duchy of Luxembourg* (C-472/98), *Commission of the European Communities v. Federal Republic of Austria* (C-475/98), *Commission of the European Communities v. Federal Republic of Germany* (C-476/98), online europa http://curia.eu.int/ (date accessed: 5 November 2002) [hereinafter ECJ judgment].

[70] The Court states that Regulation 2409/92, article 1(2)(a), only applies to fares and rates charged by Community air carriers, however, affects the air carriers of non-Member States as it 'limits the freedom of those carriers to set fares and rates where they operate on intra-Community routes by virtue of the fifth-freedom rights which they enjoy'. Moreover, as regards CRSs, the Court states that 'articles 1 and 7 of Council Regulation 2299/89 applies to nationals of non-member countries where they offer to use or use a CRSs in Community territory'. Thus, the Community acquires exclusive competence to contract with non-member countries the obligation relating to CRSs offered for use or used in its territory.

First, on the application of Article 52 of the Treaty, it applies in the field of air transport, as no Article in the Treaty precludes its provisions on freedom of establishment from applying to transport. The application of Article 52 depends, in fact, on the question whether the situation under consideration is governed by Community law. Although the alleged discrimination among Member States relates to the exercise of traffic rights to points situated in non member countries, the Court declares that 'all companies established in a Member State within the meaning of Article 52 of the Treaty are covered by that provision, even if the subject-matter of their business in that State consists in services directed towards non member countries'.[71]

Second, on the fact that the ownership and control clause is contrary to Article 52, the Court recalls that 'Article 52 guarantees nationals of Member States of the Community who have exercised their freedom of establishment and companies and firms which are assimilated to them the same treatment in the host Member State as that accorded to nationals of that Member State'.[72] Accordingly, Community airlines actually suffer discrimination which prevents them from benefiting from the treatment which the host member State grants to its nationals. The Court adds that the direct source of that discrimination is not the possible conduct of the US *vis-à-vis* the airlines of a host Member State, but the ownership and control clause of the bilateral agreements which specifically grants the right of the US to act in that way. Furthermore, the Court refuses the arguments of Luxembourg, Denmark and Finland, that consider the ownership and control clause as traditionally incorporated in bilateral air transport agreement and that, as such, they are intended to preserve the right of a non member country to grant traffic rights in its airspace only on the basis of reciprocity.[73]

Third, on the justification of the ownership and control clause on the grounds of public policy, under Article 56 of the Treaty, as argued by the UK, Germany and Denmark, the ECJ recalls that this justification 'presupposes the need to maintain a discriminatory measure in order to deal with a genuine and sufficiently serious threat affecting one of the fundamental interests of society' and 'there must be a direct link between that threat and the discriminatory measure adopted to deal with it'.[74]

The Open-Skies decision represents an essential step in the evolution of the EU external air transport policy. Ten years after the creation of the unified internal air transport market, the question of the EU dual policy, internal versus external, still prevails. However, by prohibiting the nationality clause in the bilaterals, and asking the Member States its replacement by the Community clause, the Court allows the external air transport policy to be more in conformity with the internal EU Regulations. Accordingly, with regard to the ownership and control issue, the judgment leads to more equality between EU Member States, as the community clause requirement remove the discrimination between States. With regard to

---

[71] ECJ judgment, *supra* note 69, C-476/98 para. 146.

[72] *Ibid.*, para. 148.

[73] ECJ judgment, *supra* note 69, C-469/98 para. 130.

[74] *Ibid.*, para. 57.

Community competence, although the bilateral regime will certainly not end in a bid to the ECJ judgment, as individual Member States still have the competence to conclude bilateral agreements with third countries, the judgment gives the Commission the legitimacy to start building an external air transport policy. Once the Commission gets a mandate to negotiate with third countries on behalf of the EU Member States, it will mark the beginning of a long political negotiating process, that, in the long run, will promote the EU on the international scene.[75]

## 4. The Canadian Regime

### 4.1 Canadian Law

Canadian law related to ownership and control of airlines is very similar to that of its influential neighbor to the south. Article 55 of the *Canada Transportation Act*[76] defines 'Canadian' as:

> a Canadian citizen or a permanent resident within the meaning of the Immigration Act, a government in Canada or an agent of such a government or a corporation or other entity that is incorporated or formed under the laws of Canada or a province, that is controlled in fact by Canadians and of which at least seventy-five per cent, or such lesser percentage as the Governor in Council may by regulation specify, of the voting interests are owned and controlled by Canadians.[77]

The Articles of Association of Air Canada, established pursuant to the *Air Canada Public Participation Act*[78], contain the same ownership and control requirement. Article 6(1)(b) of this Act sets forth, *inter alia*:

> Provisions imposing constraints on the issue, transfer and ownership (…) of voting shares of the Corporation to prevent non-residents from holding, beneficially owning or controlling, directly or indirectly, otherwise than by way of security only, in the aggregate voting shares to which are attached *more than twenty-five per cent of the votes* that may ordinarily be cast to elect directors of the Corporation.[79]

---

[75] For further analysis on the ECJ ruling on the Open-Skies cases, see Part 3, Chap.5, para. 4, below.

[76] *National Transportation Act*, 1987, R.S.C. 1985, c. 28 (3rd Supp.); *Canada Transportation Act, Act to Continue the National Transportation Agency as the Canadian Transportation Agency, to Consolidate and Revise the National Transportation Act, 1987, and the Railway Act and to Amend or Repel Other Acts as a Consequence*, assented to May 29th, 1996, Chapter C-10.4 [hereinafter *Canada Transportation Act*].

[77] *Canada Transportation Act, ibid.* at Article 55.

[78] *Air Canada Public Participation Act, Act to Provide for the Continuance of Air Canada under the Canada Business Corporations Act and for the Issuance and Sale of Shares thereof to the Public*, assented to August 18, 1988, Chapter A-10.1 (hereinafter *Air Canada Public Participation Act*).

[79] *Ibid.* at Article 6(1)(b) [emphasis added].

Under the *National Transportation Act*, licenses for domestic services may be issued only to a 'Canadian',[80] unless the Minister of Transport determines that it is in the public interest to grant an exemption to this requirement.[81] The same 'Canadian' requirement applies to the licensing of Canadian international scheduled and non-scheduled air services, with the important distinction that no exemptions can be granted.[82] While satisfaction of the ownership criterion can be objectively determined, based on the mandated ownership percentages, the notion of 'control' is not defined by the Act and, thus, like similar regulations in the US and other nations, it remains vague. A Report of the Standing Committee on Transport explains the notion of 'control' by simply stating that 'in determining where control "in fact" lies, the Canadian Transportation Agency analyzes financial, managerial and operational relationships.'[83]

## 4.2 Canadian Cases

One of the most important cases on the airline nationality issue, the Canadian Transport Commission (CTC) has had to deal with, was the *Okanagan Helicopters Ltd. Change of Control* case in 1983.[84] Here, United Helicopters Ltd., a British company, sought to acquire 49 per cent of the ordinary voting shares and 100 per cent of preference non-voting shares in Okanagan, a Canadian firm. The CTC rejected the transaction on the grounds that even if it were approved, it was highly unlikely that the United Kingdom would reciprocate and allow Canadian companies to hold controlling interests in UK air carriers and, therefore, allowing the Okanagan transaction would merely restrict competition and prejudice the Canadian public interest.[85]

The *Air 2000 Airlines* case similarly shows the stringent approach of Canadian regulators with respect to foreign ownership and control of airlines. In this case, the National Transportation Agency (NTA)[86] rejected an airline organizational proposal because, among other things, 25 per cent of the Canadian

---

[80] *Canada Transportation Act, supra* note 76 at Article 61.

[81] *Ibid.* at Article 62.

[82] *Ibid.* at Article 69 and Article 73.

[83] Standing Committee on Transport, *Restructuring Canada's airline industry: fostering competition and protecting the public interest*, report (December 1999), online: Canada's Parliament http://www.parl.gc.ca/InfoComDoc/36/2/TRAN/Studies/Reports/tranrp01/09-rap-e.htm (date accessed: 11 May 2001).

[84] At that time, there was no statute that specifically addressed the matter of foreign ownership and control of Canadian Airlines, as the *National Air Transportation Act* was drafted in 1987. However, the Canadian regulatory bodies (the Air Transport Board and later the Canadian Transport Commission), have required, as a matter of government policy, that Canadian air carriers be owned and effectively controlled by Canadians.

[85] Air Transport Committee, *Okanagan Helicopters Ltd. Change of* Control, Decision No. 7791 (15 December 1983); for more information on the case, see J.S. Gertler, 'Nationality of Airlines: Is It a Janus with Two (or More) Faces?' (1994) 19:1 A.A.S.L. 211 at 244.

[86] The National Transportation Agency was created by the 1987 Act as a successor of the former CTC.

airline was to be owned by Air 2000, an English carrier. In 1988, after a complete restructuring of the airline, the Agency finally gave its approval to the transaction with two conditions: (1) the NTA required the airline to change its name (it was renamed 'Canada 3000'), and (2) the company had to notify the Agency of any changes of shareholders, officers, directors, and generally of any circumstances which could result in non-compliance with the Canadian ownership and control provisions.[87]

Two other cases in which the NTA has dealt with the issue of foreign 'control' of a Canadian airline are also revealing. First, in the *Minerve Canada* case (1988), the Agency canceled Minerve's licence because 'Minerve S.S., a French company, was in an overriding position to influence the Board of Directors of Minerve Canada so as to constitute foreign control.'[88] In contrast, in the 1993 case of *Canadian Airlines International Ltd.* (CAI),[89] which involved the financial bailout of CAI by rival Air Canada, the NTA said, 'a larger and financially strong company would not necessarily gain control of a smaller and weaker company merely because of a business or equity alliance relationship.'[90]

As these few cases demonstrate, Canadian regulators generally tend to adhere to the letter of the law when it comes to the issue of foreign ownership and control of airlines. However, in those instances where there are superseding political and/or economic considerations, 'the flexibility of bilateralism... facilitate[s] arrangements whereby the designation and authorization of airlines [does] not depend on a strict application of the 'substantial ownership' and 'effective control' standards.'[91] This 'flexibility' is demonstrated, for example, by the fact that Canadian Airlines maintained its designation as a Canadian airline even though American Airlines effectively controlled Canadian, albeit with just 25 per cent ownership of the voting shares.

### 4.3 Canadian Protectionism Under Attack

In 1999, the Canadian Minister of Transport, David Collenette said, 'I remain committed to ensuring that our national transportation policy objectives are met and that we have a safe, healthy, Canadian owned and controlled air industry that

---

[87] National Transport Agency, *Air 2000 Airlines*, decision No. 239-A-1988 (12 August 1988); for further information on the case, see Gertler, *supra* note 85 at 246.

[88] National Transport Agency, *Minerve Canada*, Decision No. 618-A-1989 (6 December 1989); for further details, see Gertler, *ibid.* at 247; the decision can be found online: Canadian Transportation Agency http://www.cta-otc.gc.ca/decisions/1989/A/618-A-1989_e.html (date accessed: 4 October 2001).

[89] National Transport Agency, *Canadian Airlines International Ltd.*, Decision No. 297-A-1993 (27 May 1993); for further details, see *ibid.*; the decision can be found online: Canadian Transportation Agency http://www.cta-otc.gc.com/decisions/1989/A/618-A-1989_e.html (date accessed: 4 October 2001).

[90] *Ibid.* at 24

[91] See Gertler, *supra* note 85 at 249.

meets the needs of Canadians well into the 21st century.'[92] Mr. Collenette further made it clear that in establishing a framework for the restructuring of the airline industry, the question of whether it was in the public interest for the policy of ownership and control of airlines to remain unchanged was 'not up for discussion.'[93] Thus proposals for restructuring the Canadian airline industry through international merger and consolidation have been severely curtailed by a policy that essentially says that 'if the final compliance with these ownership and control requirements is not achieved, the proposal [must be] rejected';[94] and there are little prospects for any immediate change. Indeed, in March 2001, in a speech on the Canadian airline restructuring process that had been implemented one year earlier, Minister Collenette reiterated his firm position, stating: 'I still believe that we should continue to protect the domestic environment from increased foreign ownership and from cabotage,' especially, he said, since opening the Canadian market to foreign carriers would not benefit the domestic economy, as foreign airlines are 'only interested in providing service on the major routes for their own benefit.'[95]

Notwithstanding the Canadian government's stringent position, the National Transportation Act Review Commission, which was set up in March 1993 by the former Progressive Conservative government, has advocated a relaxation of the existing limitations on foreign ownership in Canadian airlines.[96] It stated that 'the *Chicago Convention* concept of the national carrier is being outdated rapidly by economic events,' and proposed the adoption of a 'new standard for evaluating foreign investment in Canadian aviation.'[97] A 1999 report on Canada's troubled air industry from the House of Commons Transport Committee likewise recommended increasing foreign ownership in the Canadian airline industry,[98] and was backed by presentations such as that given by University of British Columbia professors William Stanbury and Tom Ross in November 1999, which urged the Minister of Transport to change the positions he has adopted on foreign

---

[92] The Canadian Transportation Agency, New Release H100/99, 'Minister of Transport Issues Policy Framework for Restructuring of Airline Industry' (26 October 1999), online: Transport Canada http://www.tc.gc.ca/releases/nat/99_h100e.htm (date accessed: 10 May 2001) [hereinafter Restructuring of Airline Industry]; for more information about the Canadian restrictive policy on foreign direct investment, see Abeyratne *Aviation Trends*, *supra* note 4 at 361-362.

[93] Restructuring of Airline Industry, *ibid.*

[94] *Ibid.*, part named 'Backgrounder proposed legislation for Review Process for any Merger/Acquisition of Major airlines'.

[95] *Ibid.*, part named 'Speaking Notes for Transport Minister David Collenette – Airline Restructuring: One Year Later', Ottawa, Ontario 5 March 2001.

[96] See Gertler, *supra* note 85 at 222 note 26.

[97] *Ibid.* at 222.

[98] Council of Canadians, Immediate Release, 'Transport Committee Recommendations Threaten Safe, Affordable and Accessible Canadian Air Service' (December 1999), online: the Council of Canadians www.canadians.org/media/media-991208.html (date accessed: 11 May 2001).

ownership.[99] Moreover, a Report of the Standing Committee on Transport from December 1999 on the Canadian government's plan for restructuring of the airline industry, maintained that the 25 per cent foreign ownership limit was 'a regulatory barrier to entry because the industry requires a great deal of funding and there are not sufficient pools of capital within Canada to allow existing carriers to expand their operations.'[100]    In contrast with the position taken by the Ministry of Transport,[101] the Standing Committee opined that 'eliminating barriers to entry will result in a healthy, competitive airline industry with benefits for everyone.'[102]

The present status in Canada is thus that the two positions on foreign ownership restrictions for airlines are locking horns. On the one hand, the Ministry of Transport, headed by Mr. Collenette, remains steadfast in its protectionist view on foreign investment; a view that is likely to become even more firm given the current crisis in air transport and the increased concerns for national security since September 2001. On the other hand, promoters and advocates of Canadian airlines, recognizing the dire need within the industry for an influx of capital, increasingly push for the liberalization of Canadian restrictions on foreign ownership and control of airlines. Air Canada, itself, the nation's largest carrier, recently made it clear that 'it would support an increase in the foreign ownership restriction to 49 per cent from 25 per cent.'[103]

## 5.   Erosion of the 'Substantial Ownership and Effective Control' Principle

In the 1990s, a number of countries adopted new policies and amended rules regarding foreign investment in national airlines. In many cases (e.g., China,

---

[99] The Fraser Institute, Media Release, 'Avoiding the Maple Syrup Solution: Restructuring Canada's Airline Industry' (17 November 1999), online: the Fraser Institute http://www.fraserinstitute.ca/media/media_releases/1999/1999111http://www.fraserinstitute.ca/media/media_releases/1999/1999111 (date accessed: 8 September 2001); an interesting analysis named 'Does Foreign Ownership really Matter' can be found on the Stanbury and Ross document. The authors explain that it is understandable that governments want to protect their domestic firms 'in the name of nationalism'; however, the stringent position of the Minister of Transport is not justified, see T.W. Ross & W.T. Stanbury, 'Avoiding the Maple Syrup Solution: Comments on the restructuring of Canada's Airline Industry' publication (1999), online: the Fraser Institute http://www.fraserinstitute.ca/publications/pps/32/ (date accessed: 8 September 2001).

[100] Standing Committee on Transport, *supra* note 83.

[101] Restructuring of Airline Industry, *supra* note 92.

[102] Standing Committee on Transport, *supra* note 83.

[103] K. McArthur, 'Ottawa may ease airline ownership rules' *The Globe and Mail* (2 October 2001) A10. Moreover, it is interesting to note that Air Canada is not only 'protected' from major foreign investments, but as well, from any majority investments. Indeed, when Ottawa privatized Air Canada in the 1980s, it introduced a 10 per cent cap on individual share ownership to ensure that the airline was widely held. The limit has since been boosted to 15 per cent. Today, the 15 per cent cap is questioned and might be raised, as the foreign cap, see K. McArthur & S. Chase, 'Schwartz spurns Air Canada as Ottawa mulls ownership cap' *The Globe and Mail* (5 October 2001) B1.

Mexico, Peru, Australia, New Zealand, Bangladesh, etc.), existing regulations were relaxed, though for different reasons.[104] In addition to these changes to national laws, more and more deviations and exceptions to the 'substantial ownership and effective control' principle have been developed by the airline industry itself, which has grown impatient with the slow pace of statutory and regulatory changes in this area.

## 5.1 Pathways to Change

Foremost among the existing deviations from the traditional principle of national ownership and control has been the creation of multi-national airlines. These multi-nationals are normally comprised of airlines from the same geographic region, grouped together in an effort to strengthen their respective markets through a common identity. For example, Scandinavian Airlines System (SAS) is a joint operating organization of the national airlines of Norway, Sweden, and Denmark, which was created in 1951. Each of the SAS component airlines is substantially owned and controlled by nationals of the countries concerned. However, SAS is appointed as the designated airline and, thus, the holder of traffic rights in each of the three bilateral agreements concluded with third countries.[105] Other examples of multi-national airlines include Air Afrique, which was created in 1961 between eleven African States,[106] Gulf Air, which was created in 1950 between four Persian Peninsula Partner States (the national carriers of Bahrain, Oman, Qatar and Abu Dhabi (the United Arab Emirates), and Alliance Air, which was founded by the governments of Uganda and Tanzania, in conjunction with South African Airways.[107]

Another more recent deviation from the 'substantial ownership and effective control' principle can be found in a multilateral Open-Skies agreement. In November 2000, the United States and four State Parties to the Asia Pacific Economic Cooperation Agreement (Brunei, Chile, New Zealand, Singapore), joined more recently by Peru, concluded an Open-Skies Agreement,[108] that did away with the traditional requirement that an airline must also be 'substantially owned' by nationals of the designated country. While the *APEC Agreement* retains the *Bermuda I*-type requirement that an airline be 'effectively controlled' by nationals of the State whose government designates the airline to exercise traffic rights, the 'substantial ownership' requirement is replaced by a requirement that the designated airline simply be incorporated in that State and have its principal

---

[104] WTO doc.1, *supra* note 1 at 6.

[105] IATA doc., *supra* note 3 at 20.

[106] The 11 States gathered in Air Afrique are Benin, Burkina Faso, Congo, Centrafrique, Côte d'Ivoire, Tchad, Togo, Mali, Mauritanie, Niger, and Senegal.

[107] Alliance Air can be seen to be a resurrection of the defunct East African Airways; for more details about these international airlines, see IATA doc., *supra* note 3 at 20.

[108] On 15 November 2000, the five partner States reached an agreement, called *Multilateral Agreement on the Liberalization of International Air Transportation* [Hereinafter *APEC Agreement*].

place of business there. By eliminating the 'substantial ownership' requirement, this multilateral Open-Skies agreement could open the door to increased cross-border investment in national airlines that have historically been forced to rely almost exclusively on domestic sources of investment capital.[109]

Along the same lines, new criteria have been established that redefine the traditional requirement of national ownership and control and which can thus be viewed as deviations. For instance, Hong Kong uses the 'principal place of business' standard in its bilaterals. The Hong Kong clause grants the contracting party the usual right to revoke the operating permit granted to the airline designated by Hong Kong 'in any case where it is not satisfied that that airline is incorporated and has its principal place of business in Hong-Kong.'[110] Another criterion that deviates from the traditional notion of 'national carrier' is the concept of 'community of interest' that was first introduced by ICAO in 1983.[111] According to this concept, the traditional conditions for *national* ownership and control are instead employed by a group of countries that share the same regional or economic interests. For example, although British West Indies Airways (BWIA) is substantially owned and controlled by Trinidad and Tobago, other Member States in the Caribbean Economic Community (CARICOM) have designated BWIA as the carrier that receives the traffic rights granted by third countries (e.g., US, UK, and Germany) under their respective bilateral agreements.

Notwithstanding these many deviations, however, the 'substantial ownership and effective control' principle has not been effectively superceded; in fact, it remains the dominant standard for establishing the nationality of airlines, particularly as it relates to the designation of traffic rights. However, the force of these possible pathways of change is increasingly enhanced by an ever-growing number of exceptions to the principle.

## 5.2 Exceptions to the Principle of National Ownership and Control

More and more, States are amending the legal regime that govern foreign ownership of their domestic air carriers and thereby break down the time-honored system of national restrictions. Many different factors have been behind this push. For some developing countries, increased foreign investment limits were necessary in order to obtain the financing needed to keep their national air transport activity operational, if not solvent. For example, in the 1990s, Brazil raised its ceiling on foreign ownership from 20 per cent to 49.5 per cent, Korea raised its cap from 20 per cent to 49 per cent, Thailand went from 30 per cent to 49 per cent, and Peru upped its limit to 70 per cent. Bangladesh even went so far as to permit operation

---

[109] For more details about the *APEC agreement*, see IATA doc., *supra* note 3 at 21; WTO, *Note on Developments in the Air Transport sector Since the Conclusion of the Uruguay Round, Part Four*. WTO Doc. S/C/W/163/Add.3 (2001) 25 [hereinafter WTO doc.2]; D. Knibb, 'Bilateral Accord Sparks Ownership Debate...as APEC Moves Towards Multilateral Open Skies' *Airline Bus.* (January 2001) 24.

[110] IATA doc., *supra* note 3 at 24.

[111] For the 1983 *ICAO Resolution*, see IATA doc., *ibid.* at 22.

of its domestic carriers by joint ventures and unlimited foreign ownership of its cargo airlines. Then again, other countries have eliminated national restrictions altogether and have allowed 100 per cent foreign capital investment in their airlines due primarily to their geographical setting. For example, with the emergence of Singapore as a major transit hub in the Asian-Pacific region, the Singaporean government saw fit to abolish ownership restrictions that had limited foreign investment in its national airline, Singapore International Airline (SIA) to 27.5 per cent, so that foreign investors could now hold 100 per cent of SIA.[112] Likewise, both Australia and New Zealand were recently forced to accede to the reality of their isolated markets and relax limits on foreign ownership of airlines. Under the foreign acquisition regulations implemented by Australia in 1999, foreign carriers can own up to 49 per cent of an Australian international airline and 100 per cent of an Australian domestic carrier.[113] New Zealand, on the other hand, has removed foreign investment limits for its airlines altogether, but still requires that 'control' of the airline must remain with its nationals.[114]

In addition to the changes in national airline ownership regimes, there have been a number of instances where airlines have departed from the traditional ownership and control requirement, oftentimes in spite of their respective national regulations. Case in point is the relationship between Iberia and Aerolineas Argentinas. Iberia (a Spanish airline and partner, along with American Airlines and British Airways, in the Oneworld Alliance) is the biggest investor in Aerolineas Argentinas, owning more than 60 per cent of Argentinian air carrier. While Iberia's ownership of Aerolineas Argentinas is permissible under

---

[112] N. Ionides, 'Expanded Horizons' *Airline Bus.* (Nov. 1999) 34 at 36.

[113] The difference between these two percentages is due to the special protection of Quantas, viewed as an 'Australian Icon', see Knibb, 'Australian Ownership Rules Criticised' *Airline Bus.* (August 1999) 26; 'Australian Government to Ease Foreign Ownership restrictions' *Aviation Daily* (19 August 1999) 3; IATA doc., *supra* note 3 at 24; Australian Commonwealth Department of transport and regional services, *International Air Services*, Policy Statement (June 2000), online: the Australian Commonwealth Department of Transport and Regional Services http://www.dotrs.gov.au/aviation/intairservices.pdfhttp://www.dotrs.gov.au/aviation/intairse rvices.pdf (date accessed: 10 May 2001).

[114] Knibb, see *supra* note 109 at 24; moreover, Air New Zealand suffers from lack of capital, so the NZ government currently analyzes foreign bids to recapitalize the national carrier; this urgent need of capital has probably also pushed the government to relax national restrictions on ownership of airlines, see S. Bartholomeusz, 'Ansett's survival goes to the heart of deregulation policy' *The Age* (7 September 2001), online: The Age http://www.theage.com.au/business/2001/09/07/FFX2JVQF9RC.html (date accessed: 7 September 2001); G. Thomas, 'Air NZ plummet hits SIA and BIL' *The Age* (25 September 2001), online: The Age http://www.theage.com.au/news/national/2001/09/25/FFXEPDK02SC.html (date accessed: 26 September 2001); G. Evans, 'Air NZ shares bounce as talks continue' *The Age* (26 September 2001), online: The Age http://www.theage.com.au/news/national/2001/09/26/FFX44YQ02SC.html (date accessed: 26 September 2001); Z. Coleman, 'Government rescues Air NZ' *The Globe and Mail* (5 October 2001) B7.

Argentine's domestic law, which allows foreign airlines and/or other investors to own up to 70 per cent of an Argentinian airline, it is patently inconsistent with the nationality clause in the US-Argentine bilateral agreement. Nevertheless, the Argentinian government welcomed the Iberia deal for the infusion of capital it gave to Aerolineas Argentinas, while the US acquiesced to the Iberia-Aerolineas Argentinas arrangement in exchange for expanded traffic rights from the Argentinian government, arguably providing proof positive of the advantages that a liberal airline ownership regime would afford.[115]

Perhaps no case better demonstrates how cross-border investment can be used to save national airlines from bankruptcy than the joining of Swissair and Sabena. On 25 January 2001, the two airlines entered into an agreement whereby SairGroup (a subsidiary of Swissair) would increase its stake of ownership in Sabena from 49.5 to 85 per cent. The Swissair/Sabena agreement was to take effect once the bilateral agreement between Switzerland and the EU, which would make EU rules applicable to Switzerland, entered into force.[116] However, because of financial difficulties encountered by Swissair, the January accord was cancelled and a new agreement was signed on 17 July 2001, pursuant to which SairGroup ownership of Sabena would remain at 49.5 per cent. But after a financial package from the Belgian government failed to address Sabena's need for additional capital, Swissair agreed to contribute up to 60 per cent of a restructuring plan.[117] Of course, in retrospect, the wisdom of Swissair's decision to bail out Sabena is in doubt, as it likely contributed to the two carriers' recent financial collapse.[118]

## 6. Conclusion

In the global air transport marketplace, which favors full market access, concentration, competition, and multi-lateralism, bilateral agreements with their restrictions on ownership and control of airlines are no longer relevant. Nevertheless, States are slow to react to new economic trends, mainly because of protectionist considerations. While few States have taken the initiative in relaxing their regulations, airlines have not stood still waiting for change. Through cross-border investment transactions, airlines progressively become global, following the trend of many other important industries. Clearly, the international community must do more to liberalize the air transport sector and expand market access.

---

[115] H.P. van Fenema, 'Ownership Restrictions: Consequences and Steps to be Taken' (1998) 23 Air & Space L. 63 at 65 [hereinafter van Fenema 'Consequences and Steps to be taken']; IATA doc, *supra* note 3 at 25; about the need of capital of the Argentinian carrier, D. Knibb, 'Aerolineas Rescue Relies on Spain' *Airline Bus.* (August 2000) 18.

[116] IATA doc., *ibid.* at 26.

[117] Swissair Group, New Release 20/01/DC, 'Swissair Group et le gouvernement belge signent un accord sur Sabena' (17 July 2001), online : Swissair Group
http://www.swissairgroup.com/apps/media/press/index.html/?period=archive&language=f#?
period archive1language=f (date accessed: 27 September 2001).

[118] D. Michaels & R. Thurow, 'Swiss banks draw ire' *The Globe and Mail* (5 October 2001) B7.

Indeed, as we have seen, even those few liberal countries that have relaxed the national ownership requirement, often still maintain the 'control' criterion. Still, the removal of some national limitations on foreign ownership is a positive first step, since restrictions represent the biggest impediment to free circulation of capital between international industries and are a major reason for the lack of globalization of the air transport industry.

# PART 2

# JUSTIFICATIONS OF NATIONAL RESTRICTIONS REVISITED

# Introduction to Part 2

For several decades, the question of the justification for national restrictions in the international airline industry has been debated between national and international aviation entities, but absolutely no consensus has been reached by the international community. Presently, more than 20 years after the first deregulation act of the air transport sector, eight years after the last ICAO Air Transport Conference, and only a short time before the next ICAO Conference, it is time fully and objectively to assess whether legitimate reasons that favor foreign investment restrictions remain. Particularly in the light of the progressive shift of the airline industry towards a more liberal market, it must be asked whether there are credible reasons for maintaining such a restrictive regime. Chapter 3 therefore surveys the whole debate regarding the ownership restrictions, and advocates their abolishment; Chapter 4 examines the legal and economic consequences and the benefits of a regime overhaul.

# Analysis of Legal, Economic, and Security Justifications of the National Restrictions

## 1. Introduction

In the year 2000, the American Bar Association (ABA), through its Air and Space Law Forum Special Committee on Cross-Border Investment and Right of Establishment in the International Airline Industry, raised a series of pertinent issues 'in determining whether there should be any change in US laws on foreign investments in the airline industry and whether it is feasible to work towards a common international standard on cross-border investments.'[1] As most of the issues raised by the ABA are not only related to the US airline industry, but also concern the entire international airline industry, the issues will be surveyed in the following analysis.[2]

To reach a more objective conclusion on the prospective need for changing national restrictions, two important issues must be taken into account. First, what is different or unique about the airline industry, compared to other industries, that requires special restrictions on foreign investments? Due to its great military, political, and security importance, aviation has been particularly protected by States since the First World War.[3] However, times have changed since the beginning of aviation, and even though these characteristics are still true today, commercial aspects have become equally important. Thus, airlines have progressively become a real industry, facing the same competition rules as any other industries.[4] Today, very few industries are concerned by foreign investment

---

[1] American Bar Association, 'Cross-Border Investment in International Airlines: Presenting the Issues' (2000) Air & Space Law. 20 [hereinafter ABA doc.].

[2] The arguments analyzed hereunder are the ones most frequently used by governments; therefore, this list of arguments does not aim at being exhaustive.

[3] H. Wassenbergh, 'Towards Global Economic Regulation of International Air Transportation through Inter-Regional Bilateralism' The Hague (August 2001) at 5 [Unpublished].

[4] Schless states that '[o]nce the free market principles are implemented there is no reason to expect aviation to develop differently from other global industries like computers, cars, chemicals, fashion or media', see A.L. Schless, 'Open Skies: Loosening the Protectionist Grip on International Civil Aviation' (1994) 8 Emory Int'l L. Rev.435 at 461; see also R.

restrictions, and even 'strategic' industries are more and more subject to market rules.[5] Nevertheless, opinions vary with respect to the wisdom of altering the substantial ownership and effective control principle. Many people argue that some of the particularities like security and safety should never be dropped for economic reasons since the original reasons why special restrictions on foreign investment had been imposed in the sector still exist.

The second issue that must be addressed is whether national restrictions on foreign ownership and control serve the 'public interest'. The notion of 'public interest' is very broad: it includes the interest of States,[6] of airlines and their employees, of passengers, of investors and of any other contractors involved directly or indirectly in the airline industry. Accordingly, do ownership and control restrictions serve the legal and economic public interest, as well as the security public interest? Wassenbergh stresses the significance of this notion by encouraging governments to take into account the needs of society before implementing any new policy,[7] indicating that '[t]o pursue a liberal air policy (...) will require governments to recognize the international public interest as of primary importance, as their national interest will depend on increased international co-operation.'[8]

---

Doganis, 'Relaxing Airline Ownership and Investment Rules' (1996) 21 Air & Space L. 267 at 269.

[5] The US still remain very protectionist, imposing national restrictions on many sectors, such as broadcasting, electric power, nuclear power, and shipping, see S.M. Warner, 'Liberalize Open Skies : Foreign Investment and Cabotage Restrictions Keep Non Citizens in Second Class' (1993) 43 Am. U. L. Rev. 277 at 304. In Canada, national ownership and control is required in certain sectors as well; for instance, '[b]oth the *Telecommunications Act* and the *Broadcasting Act* limit foreign ownership of operating companies in the industries to 20%', see D. Johnston, D. Johnston, S. Handa, *Getting Canada Online – Understanding the Information Highway* (Toronto: Stoddart Publishing Co. Limited, 1995) 121; however, presently, Canada is pushing for ownership limit changes, see P. Brethour, 'Telecom ownership review coming: AT&T' *The Globe and Mail* (30 October 2001) B1 & B6; see also, K. Damsell, 'Ownership rules key: Astral' *The Globe and Mail* (14 December 2001) B5.

[6] The ABA raises the following questions: 'what benefits, if any, might accrue to US interests by a change in the rules pertaining to foreign investments in US airlines?' and 'are the national interests served by restrictions on foreign investment in airlines fundamentally directed to 'ownership' and 'control'?', see ABA doc., *supra* note 1 at 20.

[7] 'Government tasks are to deploy activities which are in the public interest, which fulfill an essential role in the society, which are not supposed as such to yield a profit, at least should nor be undertaken nor exploited for profit and activities that are needed by the society, but do not find a private undertaking to deploy them', see Wassenbergh, *supra* note 2 at 12.

[8] *Ibid.* at 15.

## 2. Legal and Economic Arguments

States view aviation as vital to their national economic interests and consequently feel a need to support and sustain their own airlines.[9] As a result, a number of legal and economic reasons have commonly been forwarded by governments in favor of national restrictions on foreign investments. However, the legitimacy of most of these arguments is questionable and will therefore be examined to determine the risks of the elimination of foreign investment restrictions, which would enable the international airline industry engage in mergers and takeovers.

### 2.1  The Interests of Passengers

How would the liability of an airline be determined, in case of an accident for instance, if it were majority owned by foreign citizens? This first legal argument has been raised in order to protect passengers who wished to pursue a negligent air carrier for compensation. The current legal regime is based on the *Chicago Convention*[10] and on the 'Warsaw system'[11] (on 28 May 1999, a new Convention

---

[9] Doganis, *supra* note 4 at 268.

[10] *Convention on International Civil Aviation*, 7 December 1944, 15 U.N.T.S. 295, ICAO Doc. 7300/6 [hereinafter the *Chicago Convention*].

[11] The 'Warsaw system' is composed by eight private international air law instruments, which are: the *Convention for the Unification of Certain Rules Relating to International Carriage by air*, signed at Warsaw, 12 October 1929, 137 L.N.T.S. 11, 49 Stat. 3000, T.S. 876, ICAO Doc. 601 (entered into force on 13 February 1933) (hereinafter the *Warsaw Convention*), the *Protocol to Amend the Convention for the Unification of Certain Rules Relating to International Carriage by air signed at Warsaw, 12 October 1929,* done at the Hague, 28 September 1955, 478 U.N.T.S. 371, ICAO Doc. 7632 (entered into force on 1 August 1963) (hereinafter *The Hague Protocol 1955*), the *Convention Supplementary to the Convention, for the Unification of Certain Rules Relating to International Carriage by Air Performed by a Person Other than the Contracting Carrier*, signed in Guadalajara, 18 September 1961, 500 U.N.T.S. 31, ICAO Doc. 8181 (entered into force on 1 May 1964) (hereinafter the *Guadalajara Convention 1961*), the *Protocol to Amend the Convention for the Unification of Certain Rules Relating to International Carriage by air signed at Warsaw, 12 October 1929, as Amended by Protocol done at The Hague on 28 September 1955*, signed at Guatemala City, 8 March 1971, ICAO Doc. 8932 (not yet in force) (hereinafter the *Guatemala City Protocol 1971*), *Additional Protocol N°1 to Amend the Convention for the Unification of Certain Rules Relating to International Carriage by air signed at Warsaw, 12 October 1929*, signed at Montreal, 25 September 1975, ICAO Doc. 9145, 22 I.L.M. 13 (not yet in force) (hereinafter the *Additional Protocol N°1*), the *Additional Protocol N°2 to Amend the Convention for the Unification of Certain Rules Relating to International Carriage by air signed at Warsaw, 12 October 1929*, signed at Montreal, 25 September 1975, ICAO Doc. 9146, 22 I.L.M. 13 (not yet in force) (hereinafter the *Additional Protocol N°2*), the *Additional Protocol N°3 to Amend the Convention for the Unification of Certain Rules Relating to International Carriage by air signed at Warsaw, 12 October 1929*, signed at Montreal, 25 September 1975, ICAO Doc. 9147, 22 I.L.M. 13 (not yet in force) (hereinafter the *Additional Protocol N°3*), the *Montreal Protocol N°4, to Amend the Convention for the Unification of Certain Rules Relating to International Carriage by air signed at Warsaw, 12 October 1929*, signed at Montreal, 25 September

was signed in Montreal to update the 'Warsaw system').[12] In cases of foreign ownership, this liability regime does not change, as it is an international regime applicable to the whole international community. However, there must be a clearly identifiable locus of responsibility for the safety of airlines. The concern is that if a carrier is owned by nationals who are not citizens of the designating country, it may be difficult to demonstrate the designating government's continuing competence in the technical aspects of airline and aircraft certification. In practice, this concern does not seem to be a problem because, whatever the owner's nationality, the aircraft that caused damages must be registered and the State of registration is responsible for any technical problems.[13] Therefore, if, for instance, a French air carrier is majority owned by Canadians citizens, but the aircraft is registered in France, the French State will be recognized as responsible for any technical defects of the aircraft.

Will consumers benefit from the easing of cross-border investment limitations, in terms of price and service? Consumers usually oppose international airline mergers, because they fear 'price grouping' and reduced service, but this can be shown to be groundless. First, the fear of a price increase stems from the idea that 'when two airlines merge, fares generally rise where the new airline dominates the market.'[14] This idea arose following the past merger wave between US airlines, such as when Trans World Airlines (TWA) acquired Ozark Airlines in 1986. In fact, even though these mergers have decreased competition and raise prices to some degree,[15] they occurred in the already very concentrated US market. With respect to cross-border investments in the world market, there would be completely different consequences. E. Perkins, while criticizing the mergers of airlines, has himself admitted some advantages of merging. He explains that a merger of competitors usually leads to significant economies of scale, which leads to lower prices, but this process has not worked in the US market because 'each of the seven giant airlines is large enough: getting bigger isn't going to gain much in the way of economies of scale.'[16] On a worldwide scale, on the other hand, there are still many airlines that must become stronger (through cross-border investment operations) to become more competitive and enable them to offer lower prices. It can be argued that multiple mergers in an industry-wide consolidation would even tend to decrease fares, and one of the reasons mentioned is the presence of low-fare

---

1975, ICAO Doc. 9148, 22 I.L.M. 13 (not yet in force) (hereinafter the *Additional Protocol N°4*).

[12] The *Convention for the Unification of Certain Rules for International Carriage by Air*, signed at Montreal, 28 May 1999, ICAO DCW Doc. No. 57 (not yet in force) (hereinafter the *Montreal Convention*); for more information about the Montreal Convention, see T.J. Whalen, 'The New Warsaw Convention: the Montreal Convention' (2000) 25:1 Air & Space L. 12.

[13] The *Chicago Convention*, Article 17-21, *supra* note 10.

[14] L. Miller, 'Airline Merger Offers Fliers No Pie in Sky' (1996) Wall St. J. Eur. 8 at 8.

[15] *Ibid.*; J. Mosteller, 'The Current and Future Climate of Airline Consolidation: The Possible Impact of an Alliance of Two Large Airlines and an Examination of The Proposed American Airlines-British Airways Alliance' (1999) 64 J. Air L. & Com. 575 at 600.

[16] E. Perkins, 'Mergers will squeeze consumers' (1997) Orange County (Cal.) Reg. D04.

carriers.[17] Moreover, another advantage of airline consolidation for consumers is the increased scope of loyalty programs. Indeed, passengers or any other airline consumers who participate in FFP can receive benefits because merged airlines usually have more cities offered as destinations for frequent-fliers miles.

Aside from the fear regarding prices, passengers are concerned about the reduction of service quality. Indeed, 'bumpy' service was the short-term result of the US airline consolidation, mainly because airlines had to work out problems with their schedules and operations.[18] Thus, the DOT received thousands of consumer complaints in 1987.[19] On a global basis, however, international airline mergers could benefit airline service to consumers in the long run. For example, carriers can learn more efficient and popular service techniques from each other, and can share their know-how and their operating means. In addition, as cross-border investments strengthen weaker carriers, competition might increase among the carriers. With the increase in competition, it would not be in the interest of airlines to lessen the quality of their services; if they do so, airlines would lose consumers and lose the benefit of the costs that had been cut.

## 2.2 The Interest of Airline Employees

How will the interests of labor be affected if foreign investment restrictions are relaxed?[20] The removal of ownership and control limits, which leads to international mergers, have a double effect on airline's employees.

The first effect is on the employment itself. So far, employees have always opposed any change in national regimes, for fear that it would result in a loss of jobs. Many unions are concerned that foreign investors would use their control over national airlines to replace national workers with foreign workers.[21] Others have taken the BA/USAir proposal as an example to prove the legitimacy of the risk of job loss.[22] However, none of these arguments are really justified, they just

---

[17] Mosteller, *supra* note 15 at 601.

[18] *Ibid.* at 602; Miller, *supra* note 14 at 9.

[19] 'Complaints About U.S. Airlines Up Almost Seven-Fold In August' (1987) J. Rec. (Okla. City).

[20] ABA doc., *supra* note 1 at 22.

[21] K. Bohmann, 'The Ownership and Control Requirement in U.S. and European Union Air Law and U.S. Maritime Law – Policy; Consideration; Comparison' (2001) 66 J. Air L. & Com. 689 at 713; pilots' associations have expressed their fear about the impact on employment, such as the Federation of Airline Pilots' Association (IFALPA), see 'Ownership Trend Creates Need for New Links Between States and Airlines' (June 1992) 47 ICAO J. 14; and see the Coalition of Airline Pilots Associations (CAPA) document, CAPA minutes, *Memorandum of Understanding*, APA headquarters Fortworth, Texas (9-10 February, 2000), online: CAPA
http://www.capapilots.org/Download%20Files/minutesfeb910.htm (date accessed: 14 May 2001).

[22] A. Edwards, 'Foreign Investment in the U.S. Airline Industry: Friend or Foe?' (1995) 9 Emory Int'l L.R. 595 at 636; D.T. Arlington, 'Liberalization of Restrictions on Foreign Ownership in U.S. Air Carriers: the United States must take the First Step in Aviation

represent an assessment of what may happen. In the current difficult economic climate for the airline industry, airlines need to consolidate. Mosteller states that '[t]he perception of the airlines may very well be that if they do not defend themselves via consolidation, they may not be able to stay competitive against the giant alliances that would form. Thus, if the airlines did not merge, their profits could decrease by such a margin that layoffs would happen anyway.'[23] From this point of view, we can deduce that it might be more risky for the industry's employment if airlines remain isolated and competitively weak. Moreover, we can easily add that, by principle, a competitive industry produces more output than a monopolized one. This suggests that employment levels will be enhanced if a competitive industry is fostered. Ross and Stanbury agree with this theory, and, in their discussion of the Canadian industry, they affirm that, 'while a monopoly that guarantees jobs might seem like a nice way to protect employment through a restructuring, in the long run there will be more jobs in an efficient competitive market place.'[24] The competitive marketplace can largely evolve if foreign investments between international airlines are authorized, since the infusion of capital would contribute to the development of the fleets of airlines, and consequently, of their networks, which would clearly increase the need for employees.[25]

The other fear about jobs is that foreign investments lead to the employment of foreign workers instead of national workers, due to lower labor costs in some countries. Two remarks can be made on this point. First, it cannot be ignored that this situation already exists: no international regulations prevent airlines from hiring foreign crew or administrative staff. The removal of foreign ownership restrictions could enhance such a phenomenon because mergers or financial alliances would create single entities, gathering airlines from different countries, and making foreign employment easier. Thus, this is a legitimate risk that should be taken into account as a possible consequence of the authorization of international mergers and takeovers among airlines. However, in practice, it is unlikely that cross-border investments would significantly change their current staff representation around the world. Indeed, it is hard to imagine that a carrier like Air France would replace its current crew by Korean or Mexican flight attendants. While it might save money, the public image of the airline is important to protect as well. As crew members are the closest link of the airline to its passengers, national customers would probably not wish to deal with foreign interlocutors when they choose the main national airline. Major international

---

Globalization' (1993) 59 J. Air L. & Com. 133 at 165 (this article provides the arguments of the 'Big Three' US air carriers directly in competition with USAir).

[23] Mosteller, *supra* note 15 at 599.

[24] T.W. Ross & W.T. Stanbury, 'Avoiding the Maple Syrup Solution: Comments on the restructuring of Canada's Airline Industry' publication (1999) (part named: 'Protecting Jobs is Both Inefficient and Unfair') online: the Fraser Institute
http://www.fraserinstitute.ca/publications/pps/32/ (date accessed: 8 September 2001).

[25] In order to defend the BA/USAir proposal, USAir demonstrated that the infusion of capital would 'assure the employment future of 46,000 USAir employees', see Arlington, *supra* note 22 at 166.

airlines are more eager to work on the service's quality and on their image for keeping their customers' loyalty, than to find absolutely any means to reduce costs with respect to their crew. The second remark about the risk of the reduction of employment of nationals is that it is necessary to determine whether nationals of particular countries are really concerned and whether the risk does in fact exist. As in most economic sectors, in the airline industry, labor costs much less in developing countries than in developed countries. Thus, European States and the US are concerned about employment effects, since the labor costs are quite high in these countries. This could explain why employees in the US, for instance, are so reticent about foreign investment liberalization. However, they probably have no reason to fear: American labor is not endangered by the lifting of restrictions since US labor costs are in fact quite low, at least in comparison to most of the EU countries.[26] Consequently, US jobs would not likely be endangered by higher foreign investment limits as a result of the reasonable US labor costs, nor would jobs in the EU as a result of the importance of preserving brand image.

The second effect of the removal of ownership and control limits on airline's employees is the integration of employees in a new entity. Indeed, one of the most difficult tasks facing merging airlines is how to integrate their employees, as well as how to integrate the personnel, cultures, and policies of two different organizations. This issue raises the particular question of the seniority rankings, which are used by the airline pilots to determine which pilots fly the popular schedules and routes. Merging the pilot lists of two airlines could push pilots down the seniority ladder.[27] It is certainly a delicate process to adapt an airline structure to another one: it is time consuming, especially considering the seniority system of such employees as pilots, who are generally resistant to any change in their position in the hierarchy after years of service. Still, this point should not justify blocking the liberalization of foreign investment. The synergy of two organizations is possible to implement even if, for instance, the synergy of pilots would require a great deal of effort with respect to the reorganization of operations. Moreover, the removal of foreign investment restrictions does not automatically entail mergers between air carriers: it could simply lead to the contribution of capital among airlines without any need to integrate the seniority rankings of pilots.

## 2.3 The Interests of Other Involved Contractors

What are the economic costs and benefits to the national economy of relaxing restrictions on foreign ownership?[28] To what extent should national economy be concerned by the issue of the ownership and control of airlines?

Airlines contribute to their local economies in many important ways. If an airline meets financial difficulties, its activity and number of employees might be

---

[26] Accordingly, some go as far as saying that US carriers would have an economic incentive to employ US citizens rather than Europeans, Bohmann, *supra* note 21 at 714.

[27] Mosteller, *supra* note 15 at 599; McKinsey & Company, 'Making Mergers Work' *Airline Bus.* (June 2001) 110 at 111.

[28] ABA doc., *supra* note 1 at 20.

reduced. In the long run, the carrier may even disappear if it does not receive financial support. This support may come in through cross-border investments.[29] Thus, lifting foreign investment limits could help to preserve the economic well-being of airlines, and, in turn, the economy that is dependent upon them.

The current global market has an interdependent nature. It is interdependent inside the aviation industry itself where international airlines are all economically related because they all depend on the aviation market. The market, taken as a whole, is interdependent as well, as every economic activity depends on each other. In his article on foreign ownership, Arlington describes the important role that USAir has had on the Pennsylvania economy by being 'the second largest private employer in the southwestern part of the State' and 'important to many of the other local economies that it serves' in the short and long run.[30] USAir served the building trade by making enormous investments in a huge new terminal and a large hub; it employs thousands of people who work, live and contribute to the local economies all around the US; and it brings a flow of professional visitors to Pennsylvania and also enhances tourism. Indeed, tourism is just one example of an industry that is dependant on the economic well-being of air carriers. Furthermore, having a good international airline has a huge impact on the promotion of a country as an international tourism destination, in addition to all the industries that benefit from tourism. Thus, restrictions on ownership and control of airlines have two negative consequences on the tourism industry. First, if a national carrier has difficulties because it suffers from a lack of capital resources and goes bankrupt, the national tourist economy is also at risk, as would be the overall economy if tourism represents one of the main income sources for the country. The second negative effect that foreign investment restrictions have on the tourism sector is that restrictions are inherently biased against growth, as they tend to reduce the availability of new service opportunities to the level acceptable to the least competitive airline. Airlines are prevented from expanding their services at the rate they feel the market will sustain.[31] Indeed, if foreign investors were allowed to offset the lack of capital felt by airlines, then airlines could operate more efficiently. Thus, restrictions on foreign investment penalize not only air carriers, but also travelers, shippers and the overall vigor of the world economy. G. Lipman, President of the World Travel & Tourism Council, stated a few years ago that 'the airline sector must be liberated from its bilateral straitjacket' and added that 'the system's ethos of growth within restraints can no longer accommodate efficiently the growing globalization of markets, and their increasing interdependence.'[32]

---

[29] In the next sub-paragraph, we will examine why the airline industry needs more and more foreign capital to support its infrastructure and, hence, the urgency to drop foreign ownership restrictions in the airline industry.

[30] Arlington, *supra* note 22 at 169.

[31] Considering that airline services do not usually answer all the demand, as most of the agreements regulate strictly the level of services/capacity both designated national airlines are allowed to provide.

[32] G. Lipman, 'Multilateral Liberalization – The Travel and Tourism Dimension' (1994) 19 Air & Space L. 152 at 152.

Presently, the promotion of tourism is a main argument used by some governments that are willing to lift the foreign ownership limit.[33] Indeed, the current air transport recession has contributed to the tourism recession. It is, therefore, very important to modify the current restrictive bilateral system that is directly biased against growth. The public interest depends on this shift.

## 2.4 National Airline Interests

Airlines are, of course, the entities that are the most concerned by the issue of ownership and control. This question is quite controversial with respect to the interests of the airlines. On the one hand, it has been admitted by most of the international community that national airlines need outside capital, not only as a prerequisite for growth, but also as a condition for survival; hence, the necessity to remove foreign investment restrictions. On the other hand, a number of risks have often been mentioned with respect to the possible impact on airlines if they were not protected by national restrictions anymore. After studying the question of the need for outside capital resources, below, the extent to which it is risky for airlines to open their capital and control to foreigners will be analyzed. In other words, the legitimacy of the economic risks faced by the airlines will be examined.

*2.4.1 Need for outside capital in national airlines* Why is access to foreign capitals so important for the airline industry? The main idea is that foreign investments spur national economic growth and development since they bring new scarce capital resources. They represent an engine for growth in developed countries and even more so in developing countries. Indeed, the success of newly industrialized States such as Hong Kong, Singapore, and Taiwan, as well as the growth of some west African countries, is due to a 'pro'-cross-border investments policy. Foreign capital resources support national economies and benefit individual national enterprises since they increase trade and create jobs.[34]

In the airline industry, foreign capital resources have the same benefit. They help airlines remain competitive in the worldwide market, and are essential for airlines to avoid bankruptcy. Indeed, the airline industry is very fragile. It suffers from severe business risks as there are high fixed costs, highly cyclical demands, and intensive competition. Consequently, to be profitable, the airline industry should earn more than other industries; however, airlines earn less.[35] Therefore, airlines require national and foreign capital to finance their investments and expenses. Two alternatives exist for raising capital: first, internal alternatives such as asset sales or trading labor for equity; and, second, external alternatives, which

---

[33] 'ANZ Asks Government To Lift Foreign Ownership Limits' *Aviation Daily* 345:10 (16 July 2001) 5.

[34] S.S. Haghighi, *A Proposal for an Agreement on Investment in the Framework of the World Trade Organization* (L.L.M. Thesis, Institute of Comparative Law, McGill University 1999) [unpublished] at 6 and 22.

[35] P.S. Dempsey, 'Airlines in Turbulence: Strategies for Survival' (1995) 23:15 Transp. L. J. 15 at 21 [hereinafter Dempsey 'Airlines in Turbulence'].

are either national, such as government assistance, or international, with foreign investment.[36] Often, the lack of national capital resources must be offset by foreign capital contributions for national airlines to be strengthened.[37] According to this logic, foreign investments should not be limited if they should be implemented to their full use in the global economic world. Especially in the airline industry, national governments should not snub outside capital: it inhibits the proper development of air services, whereas it could enable airlines to survive or grow. The evolution of the US airline industry perfectly illustrates the need to lift national restrictions on foreign investments and open the market to unlimited outside capital resources. Following deregulation and in the 1990s, many US air carriers that faced difficulties, such as TWA or Pan Am, could have been more effectively supported by foreign capital and would have had a better chance of long-term success.[38] In addition, these foreign contributions might have allowed certain US airlines to survive.

Thus, a pertinent question is whether the airline industry needs to be supported by foreign investors. Capital needs are presently very strong. First, air carrier activity requires more and more capital contributions to finance huge expenditures – since the 1960s, capital spending by the world's airlines has continuously increased.[39] Second, during the year 2001, the air transport sector entered into a deep and medium or long-term financial crisis, which was made even worse in September. Consequently, many carriers are looking at how they can survive,[40] and some governments even ask for a more liberal interpretation of the ownership and control provisions in the bilaterals 'to facilitate any cross-border mergers or acquisitions necessary to maintain the viability of the aviation industry during the crisis.'[41] Foreign capital resources are not only necessary, but are urgently required for a number of airlines, and all the world's regions are concerned by the crisis. Swissair and Sabena are recent examples of bankrupted aviation companies. Swissair's collapse, which occurred at the beginning of October 2001, was mainly due to the excessive ambition of the managers who stretched the airline beyond its financial capacity.[42] The disastrous financial outcome of the Swiss group led to the bankruptcy of the French airlines AOM and

---

[36] *Ibid.* at 78.

[37] P.P.C. Haanappel, 'Airline Ownership and Control and Some Related Matters' (2001) 26-2 Air & Space L. 90 at 96 [hereinafter Haanappel 'Airline Ownership and Control'].

[38] T.D. Grant, 'Foreign Takeovers of United States Airlines: Free Trade Process, Problems and Progress' (1994) 31 Harv. J. Legis. 63 at 71; Bohmann, note 21 at 711; Edwards, note 22 at 619.

[39] According to a 1990s study, world's airlines need about $815 billion today, compared with $147 ten years ago, see Dempsey 'Airlines in Turbulence', *supra* note 35 at 76.

[40] '1944 and all that' *Sunday Times – London* (7 October 2001), available on WL 27457432.

[41] ICAO, Working Paper (*Substantial Ownership and Effective Control over Designated Airlines*) No. A33-WP/181 (25 September 2001) [hereinafter ICAO Working Paper No. A33-WP/181]; 'UK says airline merger and acquisition rules should be relaxed' (3 October 2001) *Airline Indus. Info.*.

[42] D. Michaels & R. Thurow, 'Swiss banks draw ire' *The Globe and Mail* (5 October 2001) B7.

Air Liberté as well, after their struggles of half of the year, during which time they had requested more capital resources from their main investor, the SairGroup.[43] The Belgium airline, Sabena, also suffered, since the Swiss industry was its main investor[44] and as a result, Sabena collapsed in November 2001.[45] In North America as well, the airline industry has not escaped the current worldwide crisis, and therefore it needs capital more than ever.[46] However, national restrictions on ownership and control are so high in the US and in Canada, that, despite the outside capital need to support their activity, American and Canadian airlines are almost entirely dependent upon national resources.[47] In addition to the problems faced by airlines in the most developed countries, air carriers from developing countries suffer from under-capitalization as well. The current crisis worsens the lack of national capital sources that these countries have always known.[48]

At present, only a few governments seem to understand the importance of opening their national carriers' capital to foreign investors, which would enable airlines to face the current air transport crisis and ensure their long-term survival. The New Zealand government, for instance, has finally relaxed foreign ownership limits in Air New Zealand, after a few months of debate.[49]

---

[43] For more information about AOM and Air Liberté struggle, see Mallet, 'Companies & finance international: French airlines in crisis' *Financial Times* (9 April 2001), online: Financial Times http://specials.ft.com/ln/ftsurveys/industry/sc22356.htm (date accessed: 16 May 2001); 'L'Etat est contraint de jouer au 'pompier social'' *Le Monde* (18 October 2001), online: Le Monde http://www.lemonde.fr/rech_art/0,5987,235581,00.html (date accessed: 5 November 2001); 'Air Liberté espère revenir à l'équilibre en 2003' *Le Figaro* (2 November 2001), online: Le Figaro
http://www.lefigaro.fr/cgi.bin/gx.cgi/AppLogic+FTContentServer?pagename=FutureTense/Apps/Xcelerate (date accessed: 5 November 2001).
[44] D. Michaels & R. Thurow, *supra* note 42 at B7.
[45] B. Crols, 'Flag-carrier Sabena in death spiral' *The Globe and Mail* (6 November 2001) B13.
[46] McCartney, 'Widening losses at airlines make shake up unavoidable' *The Globe and Mail* (6 November 2001) B15.
[47] K. McArthur, 'Air Canada courting investors' *The Globe and Mail* (3 November 2001) B1.
[48] For instance, in the Caribbean region, see Caribbean Alpa, *The problem with all Caribbean carriers is undercapitalization*, publication, online: Caribbean Alpa http://www.caribbeanalpa.com/discussion/posts/1546.html (date accessed: 14 May 2001); in addition, some of the Caribbean airlines suffered from the 11 September US tragedy and, as a result, had to find short term funding, see 'Air Jamaica hurt by US tragedy' *The St. Vincent Herald* (4 November 2001), online: Caribbean Alpa
http://www.caribbeanalpa.com/news/index.shtml (date accessed: 5 November 2001);
moreover, small countries, such as El Salvador, opposes national restrictions because of the need of outside capital, see H.E. Kass, 'Cabotage and Control: Bringing U.S. Aviation Policy into the Jet Age' (1994) 26 Case W. Res. J. Int'l L. 143 at 150.
[49] S. Bartholomeusz, 'Ansett's survival goes to the heart of deregulation policy' *The Age* (7 September 2001), online: The Age
http://www.theage.com.au/business/2001/09/07/FFX2JVQF9RC.html (date accessed: 7 September 2001); 'The New Zealand Government Relaxes Foreign Ownership Limits in Air New Zealand' *Air Transport World* (1 October 2001) 12.

*2.4.2 Potential risk of inhibiting transport operations*  To what extent would the air transport operations of airlines be affected if foreign investment restrictions were relaxed? The air transport operations of airlines have two components: the internal operations that include the long-term flight schedule design, decided by the planning service; and the outcome of the internal operations, that includes dealing with the possible material inconveniences resulting from the flight, such as delays, cancellations, and lost baggage.

Some have argued that airline mergers, fostered by cross-border investment liberalization, would impede the flight schedule management, and that the loss of time that would result from trying to combine two airline programs would be very costly. In addition, airlines would likely be unable to afford these costs, as '85 per cent of an airline's cost structure is fixed to its schedule.'[50] This position is understandable, given that changes to flight schedules require months of advanced planning. As the main activity of airlines, scheduling is an important and expensive process with respect to time, employees, and money. Where there is a merger, the flight schedules of the merged airlines would have to be combined, which makes the process even more complex and costly. Planning services must work on the possible program synergies between the two carriers. However, if two carriers decide to merge, it is for all the economic benefits they may eventually obtain. To assess all the benefits of such a financial strategy, airlines must maintain a long-term perspective. Thus, materially, an integrated planning department must be created to combine schedules; while financially, the new department should develop common market-profitability measures, since a large part of the synergy in an airline merger comes from network optimization.[51] To agree on solutions, common measures must be implemented. With such a well-managed policy, an airline merger can be, in the long-run, a very high-profitable operation.

Concerning the daily consequences of internal operation, such as delays, cancellations, lost baggage, or long lines, it has been argued that '[e]ven in the best times, the "product" has a high service failure rate. (…) Failing to adequately plan for operational transition has been the undoing of many airline mergers.'[52] However, mergers are not entirely to blame for service failures. First, airlines are often considered responsible for the inconveniences of air transport operations, whereas inconveniences are often caused by other entities: delays and cancellations can be due to air traffic control or airport inefficiency, lost baggage and long lines are mainly due to airport disorder or unreliable airline/airport subcontractors. Second, concerning the few failures caused by the airline's negligence, there would

---

[50] On the argument about the effects of mergers on the schedules of airlines, see McKinsey & Company, *supra* note 27 at 111 ('The general rule in successful post-merger management is that it is critical to get value quickly. But in airline mergers, a large portion of the synergies are revenue improvements, especially given the constraints in reducing cost. Unfortunately, these benefits require long-lead time decisions supported by careful planning. By the time the dust settles on the merger, these critical decisions are often far, far behind schedule').
[51] *Ibid.* at 113.
[52] *Ibid.* at 111.

seem to be no reason for greater service failures that would be directly caused by national or international airline mergers. In this area as well, therefore, the full integration of the main services are required, using such planning services as described above, and demanding an organizational and rational synergy. No trouble should be faced by either the carrier or the consumer if the merger leadership team outlines a process that not only matches its capabilities, but can also be executed smoothly.

To sum up the issue regarding the air transport operations of airlines, the relaxation of foreign investment restrictions does not inhibit the normal course of the airline's daily operations. Even in the case of mergers or takeovers, air transport operations should not be affected if the concerned airlines have a managed and monitored rational integration.

*2.4.3 Possible risk of a decline in competition*  To what extent can competition between airlines be affected if foreign investment restrictions are relaxed?[53] It is generally agreed that the lifting of foreign investment restrictions would increase competition in the airline industry; as a result, the airlines as well as the public community will benefit from this heightened competition. Indeed, the advantages of a liberal ownership and control regime with respect to competition are demonstrable in two ways. First, restrictions on the ownership and control of airlines impede new entrants from penetrating national markets. In the 1999 report of the Canadian Standing Committee on Transport, it is stated that 'raising the foreign ownership limit to a higher level would remove a significant barrier to entry and enhance competition in the domestic air market.'[54] National competition is effectively disadvantaged by all the restrictions since the enormous amount of capital needed to create new carriers may not be available in the hands of national individuals or entities.[55] Therefore, in most countries, despite the air transport liberalization, a main air carrier usually dominates the national market, and a small number of small carriers may attempt to survive alongside.[56] If national restrictions are lifted, the enhanced competition would provide lower fares and more choices for customers,[57] and would thus benefit the public interest. In fact, it would be profitable to the market as a whole, including the airline industry. To what extent

---

[53] In other words, '[t]he debate finally deals with the issue of whether the current regime imposes an obstacle to the development of liberalized and open aviation markets or whether the protection of the domestic aviation industry from foreign competition is still necessary', see Bohmann, *supra* note21 at 715.

[54] Standing Committee on Transport, *Restructuring Canada's airline industry: fostering competition and protecting the public interest*, report (December 1999), online: Canada's Parliament      http://www.parl.gc.ca/InfoComDoc/36/2/TRAN/Studies/Reports/tranrp01/09-rap-e.htm (date accessed: 11 May 2001).

[55] *Ibid.* ('[T]he industry requires a great deal of funding and there are not sufficient pools of capital within Canada to allow new entrants to compete in the market').

[56] This is the case in Canada, increasing the foreign ownership limit may ensure that Canada has two competing national airlines, rather than one dominant carrier, see Standing Committee on Transport, *supra* note 54.

[57] Bohmann, *supra* note 21 at 715.

would more competition, fostered by the withdrawal of ownership and control restrictions benefit airlines? The basic theory is that since, by definition, a monopolist has no competitors, it has no incentive to search for ways to lower its production costs. Rather, it can simply pass cost increases on to consumers in the form of higher prices. According to Hill, 'The net result is that a monopolist is likely to become increasingly inefficient, producing high-priced, low-quality goods, while society suffers as a consequence.'[58] Of course, this theory also applies to the airline industry: demand will increase only if airlines try to improve their services, both qualitatively and quantitatively. Thus, the role of government in the market economy is to encourage vigorous competition among product or service companies; in doing so, it fosters private ownership and foreign ownership.

The second advantage of the removal of foreign investment restrictions with respect to competition is the important capital contributions that could follow as a consequence. While new entrants increase competition in the domestic market, access to foreign capital enhances competition in the international market. Indeed, additional investment in the domestic airline industry would enable national carriers to compete more effectively with foreign carriers, as funding would allow existing carriers to expand their operations.[59] Access to foreign capital would enhance competition among air-faring States, as well as between developed and developing States, as it would allow the latter to reap the benefits of market competition and to compete much more fairly with dominant nations in the air carrier market.[60]

Despite the recognition of the advantages that the removal of foreign investment restrictions would have for competition, some authors insist that increased levels of foreign investment would lead to unfair competition. They argue that cross-border investments would lead to the domination of State-owned foreign airlines in the international airline industry. As a big concern for American carriers, this fear is cited by many authors: 'heavily subsidized or even State-owned foreign airlines could invest in US carriers and, due to their State financial support, enjoy a competitive advantage over other US carriers.'[61] Indeed, it is probably easier for an airline to depend on State subsidies than to seek private

---

[58] C.W.L. Hill, *International Business: Competing in the Global Marketplace*, 2nd ed. (Chicago: Richard D. Irwin, 1997) at 39 [hereinafter Hill 1997].

[59] This argument has often been used for US airlines to remain competitive on the international scene, see Bohmann, *supra* note 21 at 715; see also Department of Transportation, 'Entry and Competition in the U.S. Airline Industry: Issues and Opportunities' Special Report 255 (30 July 1999), online: DOT

http://www.ostpxweb.dot.gov/aviation/domau/dottrbre.pdf (date accessed: 10 May 2001). The same argument is used to promote Air New Zealand in the international airline market, see 'ANZ Asks Government To Lift Foreign Ownership Limits' *Aviation Daily* 345:10 (16 July 2001) 5.

[60] G.L.H. Goo, 'Deregulation and Liberalization of Air Transport in the Pacific Rim: Are They Ready for America's 'Open Skies?' (1996) 18 U. Haw. L. Rev. 541 at 562.

[61] Bohmann, *supra* note 21 at 715; Edwards, *supra* note 22 at 633; the ABA has raised this issue too by asking '[d]oes it matter whether the foreign investor is a government-subsidized airline?', see ABA doc., *supra* note 1 at 21.

capital funds. Moreover, the airlines, constantly provided with government subsidies, may be more tempted to offer lower prices to consumers than private carriers would be, and, consequently, they would dominate the market. A fair concern to raise would therefore be whether all these arguments against liberalization of cross-border investments are relevant. In fact, these arguments were probably more valid before the 1990s, when most of the airlines were still public-owned, than they are today.[62] In her article, Edwards addresses a catastrophic scenario resulting from this 'imagined' State-owned airline domination,[63] but it makes less sense now since the aviation landscape has been different for many years. Considering itself to be the most concerned State, due to its current aviation supremacy, the US has little reason for reacting against cross-border investment liberalization by criticizing the domination of State ownership of airlines. Several reasons explain why this reaction would be overstated. First, the move towards the partial or full privatization of public air carriers worldwide has made much progress in the past year, and only few important airlines remain State-owned.[64] Second, the US airline market is already so concentrated and competitive that it is unlikely that a public airline would try to penetrate the US market. Third, public airlines would not seem to be a threat because regulations with respect to State aids are becoming more and more stringent.[65] The partially privatized airlines or the subsidized airlines are increasingly limited in capital funds, so they cannot afford to either invest substantially in foreign airlines, or drop ticket fares to remain competitive. Finally, if nonetheless the US or other countries remain concerned by this risk of unfair competition, they may always limit the ownership liberalization to carriers that are not owned or controlled by their governments.[66]

Considering all the above, it can be concluded that the relaxation of foreign investment restrictions will certainly affect positively competition in the airline industry. It will benefit the public interest, as well as international air carriers.

---

[62] 'From 1985 to 1994, the governments announced privatization plans or expressed their intentions of privatization for approximately 115 national airlines. Since 1995, another 50 carriers have joined the list', see ICAO Doc. AT/122 (9 October 2001) at 2.5 [hereinafter ICAO Doc. AT/122].

[63] 'U.S. carriers will be forced to drop ticket fares to match the State subsidized carrier. (...) The strong carriers will be weakened by this artificial competition. (...) Additionally, it may prove destructive to the nationwide aviation infrastructure (...)', see Edwards, *supra* note 22 at 633-634.

[64] On 29 July 2002, the privatization of Air France has been announced officially. The part of the capital owned by the French State, currently 54.4 per cent, will be reduced up to 25 per cent, see Air France, Press Release, 'Lancement du processus de libéralisation d'Air France' (29 July 2002), online Air France http://bv.airfrance.fr/cgi-bin/FR/frameset.jsp (date accessed: 20 September 2002).

[65] For instance, with respect to the EU, see Bohmann, *supra* note 21 at 716; see also P.S. Dempsey, 'Competition in the Air: European Union Regulation of Commercial Aviation' (2001) 66 J. Air L. & Com. 979 at 1124-1139 [hereinafter Dempsey 'Competition in the Air'].

[66] Bohmann, *ibid.*.

*2.4.4 Potential risk of loss of traffic rights and the cabotage concern*   To what extent are traffic rights of national airlines linked to the ownership and control issue? This main issue will be analyzed on the basis of two important questions that represent ones of the biggest economic concerns for governments and the airline industry.

The first relevant question to be answered is whether the international traffic rights of national airlines are really endangered by the relaxation of foreign investment restrictions.

International traffic rights involve the access to routes between two States and are negotiated by bilateral agreements. State parties to these agreements designate the national airlines that are authorized to operate between the two countries and the main condition in the designation clause is the national ownership and control requirement of the designated airlines. If the lifting of foreign investment limits is decided by one State, the consequence on traffic rights will be as follows: the State A liberalizes its ownership and control law, and its airline A' is bought by B', the airline of State B. This acquisition allows B' to use the traffic rights of A', which had been previously negotiated between A and a third country C. The problem here is that C has not negotiated with B the traffic rights linking State A to State C, and there is no reason that State B should be able to access the State C' market, if C does not receive reciprocal market access into the State B.[67] There may be many reasons why C would not want B' to benefit from the traffic rights. For instance, C' would have to face more competition than foreseen in the negotiated agreements because, through financial operations with airlines of State parties to the agreements (e.g. takeovers, mergers, share purchases) third countries could benefit from the same routes. Accordingly, to protect its national economy and air carrier, C could cancel the traffic rights previously granted to A by using its revocation right granted in the designation clause of the bilateral agreement. The countries that are most subject to the risk of the loss of their traffic rights are the EU Member States because cross-border investments have been fully liberalized among the Member States, but third countries do not yet recognize the resulting 'Community carrier' status of EU airlines. Therefore, if a European airline purchases an airline from another European country, the State of the purchased airline will probably need to

---

[67] van Fenema gives some examples of this complex situation. For instance, '[i]f the US has a restrictive bilateral with France and a liberal bilateral with the Netherlands, Air France should not be able to profit from KLM's free market access into the US by buying a controlling interest in KLM. Because Air France would thus get market access into the US without having given reciprocal market access into France to the US carriers. Unthinkable in traditional *quid-pro-quo* thinking!' and then raises an interesting issue by asking 'whether the outcome, i.e. the US revoking or restricting KLM's traffic rights, would or should be different if not Air France, but an independent French tour operator, bank or private investor – neither directly nor indirectly related to Air France – would have assumed control of KLM', see H.P. van Fenema, 'Ownership Restrictions: Consequences and Steps to be Taken' (1998) 23 Air & Space L. 63 at 64 [hereinafter van Fenema 'Consequences and Steps to be taken'].

renegotiate its traffic rights with the third countries concerned.[68] This situation occurred when British Airways purchased the French airline Air Liberté; the Moroccan State imposed guarantees on the French State so that the traffic rights granted to Air Liberté on Morocco could be maintained. Outside the EU, this problem with respect to traffic rights was faced by *Aerolineas Argentinas*, which held traffic rights on the US; when *Iberia* purchased majority shares of the Argentinian airline, the US undertook a re-examination of the traffic rights. Accordingly, it seems that traffic rights of national airlines are jeopardized by the withdrawal of foreign investment restrictions. However, in order to avoid risking the loss of traffic rights, the international community could first liberalize traffic rights before starting the process of cross-border investment liberalization. Indeed, if air routes were not negotiated on a bilateral basis anymore, States could step forward in the process of liberalization and replace Open-Skies agreements by plurilateral or multilateral arrangements; this would fully liberalize air transport market. In such a context, there would be less concerns about the risk that governments and the airline industry could lose the traffic rights that are already in force in the event of the allowance of cross-border investment.[69]

The second important question to answer is: what would be the national economic benefits/costs for the opening of cabotage rights as a consequence of the liberalization of the ownership and control of airlines?

Cabotage (i.e. domestic air traffic rights) is 'the transportation of passengers, cargo, or mail by a foreign airline between two points in the same nation – the foreign carriage of domestic traffic.'[70] Article 7 of the *Chicago Convention* has affirmed the State's right to restrict cabotage.[71] Indeed, there has been a great deal of controversy over its interpretation.[72] In fact, some States have been particularly stringent about granting cabotage rights to foreign airlines,

---

[68] In that respect, van Fenema addresses the problem of the necessary renegotiation of all pertinent bilateral agreements concluded by the two concerned governments when the designated national carriers of their respective governments intend to merge: 'bilateral agreements do provide for the possibility to replace one national designated carrier by another of the same nationality (...) [b]ut an airline of "dual nationality" replacing one of single nationality in a bilateral relationship is quite another thing, a novum, not covered by the standard texts', see H.P., van Fenema, 'Substantial Ownership and Effective Control as Airpolitical Criteria' (1992) Air & Space Law: De Lege Ferenda (Liber Amicorum Henri Wassenbergh), Masson-Zwaan and Mendes de Leon eds., pp. 27-41, the Netherlands; see also H.P. van Fenema, 'Airline Ownership and Control: Long and Short Term Approaches to a Trade Barrier' (Annual Conference of the European Air Law Association, Zurich, 9 November 2001 - to be published in 2003 by Sakkoulas and Kluwer) 4-5.

[69] This issue of traffic right liberalization as a first step, to be followed by a relaxation of the foreign investment restrictions, will be further discussed in Part 3, Chap. 5, para. 2, below.

[70] Schless, *supra* note 4 at 452, note 113.

[71] The *Chicago Convention*, Article 7, *supra* note 10.

[72] The terms 'specifically' and 'exclusive basis' in the second sentence of the text have been interpreted differently, and need to be clarified. Mainly, it raises the question as to whether cabotage is granted to one State, or whether the same rights would need to be made available to other states, see Kass, *supra* note 48 at 152; and see Warner, *supra* note 5 at 314-315.

especially the US, which reject in any event that foreigners penetrate the US market, except in case of emergency.[73] The issues of cabotage and the ownership and control of airlines both raise the same national concern: the access of foreign carriers to national air transport markets. The main concern is that the liberalization of only one of these two restrictions would grant access to foreign carriers. Accordingly, the liberalization of ownership and control restrictions would be sufficient to give foreign airlines the chance to penetrate the market, and as a result, cabotage restrictions would be rendered meaningless. National governments have been wary of the competitive advantage a foreign airline could secure if given the opportunity to buy national airlines. Such an opportunity would allow foreign airlines to circumvent cabotage rules by operating within the internal air transport market through their national subsidies. Foreign ownership is thus considered, by Dempsey, as the 'back door to cabotage.'[74] This national concern has inspired governments to protect their domestic market and their airlines to a great extent, hence the remaining strict regime on foreign investments. Thus, the US has always considered foreign investors as a threat to the US airline market,[75] as, through cross-border investments, they get the opportunity to assume all the domestic routes and market advantages of the target carrier.[76]

This potential threat to the national airline industry resulting from the granting of cabotage has seen a great deal of debate. Most authors advocate the denial of cabotage rights to foreign air carriers since cabotage represents a threat to the national economy. The same arguments as those used against the increase of foreign investment are usually advanced against the granting of cabotage rights as well. These reasons include the protection of national security and of the national carrier's competitive advantage in the national market, which would also tend to safeguard national employment. However, at the present time, authors support the idea of eliminating cabotage restrictions, citing two reasons. First, the main justifications for cabotage restrictions, such as national security, are highly debatable.[77] Second, other justifications can be interpreted differently, so the current consequences of granting cabotage rights to foreign carriers would in fact

---

[73] Like for the ownership and control issue, the US maintains a very strict position on cabotage, remaining very protectionist. However, DOT's policy is evolving towards a more liberal interpretation of cabotage, see Kass, *supra* note 48 at 156-163.

[74] Edwards, *supra* note 22 at 626

[75] The US, more than any other States, feel threatened by this issue because, as it has the biggest and strongest domestic air transport market in the world, it does not want to lose its supremacy.

[76] Edwards even states that a simple increase of ownership limits to 49 per cent would give too much control to foreign airlines because this increase 'will allow foreign carriers to indirectly commit cabotage via actual control of US airlines. Therefore, liberalization of the US domestic market will create the threat of potential foreign dominance', see Edwards, *supra* note 22 at 628-629.

[77] For the ones who believe in the fact that cabotage would be too risky for national security, it can be argued that cabotage rights granted to foreign airlines can always be suspended in case of emergency or war, see Warner, *supra* note 5 at 317.

benefit national economy.[78] Thus, it is difficult to give a clear and definitive answer about the national economic consequences of cabotage. To determine whether cabotage is worthwhile for the economy, a deeper analysis of the issue would be necessary. However, we can already state that like with the ownership issue, the changing scene of the aviation industry of the past 20 years has required an evolution of the main air transport concepts and restrictions in order to maintain the competitiveness of national airline industries. Indeed, I believe that the liberalization of cabotage would result in more financial and economic benefits than economic losses.

In fact, whether the effects of cabotage on national economies are positive or not, the airline industry will probably not suffer the effects. Indeed, it would be absurd and irrational that a foreign airline sets up a route network in a country where almost every airline suffers from 'chronic economic anemia'.[79] Indeed, in the US, the airports are already surcharged, the slots are all taken, and the internal market is highly competitive, even among the very few major US air carriers. Accordingly, it would not be worthwhile for a foreign carrier to penetrate the US market.[80] On this point, Scocozza indicates that 'cabotage rights must be valued in light of the highly competitive US market, rather than on the vast size of the market.'[81] In fact, such a closed market exists in each EU Member State as well. Indeed, by the third package of liberalization,[82] cabotage between all the Member States has been decided, and it entered into effect after a transitional period which ended on 1 April 1997. However, since 1997, European airlines have not been very eager to penetrate each other's internal markets. Although the main reason is because of the risk of losing traffic rights, these hesitant initiatives are also due to the fact that each market is already very competitive and does not provide enough demand for new entrants. Moreover, opening cabotage would not significantly affect the US airline industry because the US industry is the strongest in the world[83] and, as such, has a tremendous advantage over the worldwide scale. If a

---

[78] Bliss answers positively the question '[d]o cabotage rights in fact constitute great economic opportunity?'. He states that cabotage would enhance competition in the airline industry and therefore it would benefit the industry itself and the consumer, see F.A. Bliss, 'Rethinking Restrictions on Cabotage: Moving to Free Trade in Passenger Aviation' (1994) 17 Suffolk Transnat'l L. Rev. 382 at 399, and notes 63 and 65. S. Warner develops the same idea by stressing the advantage the US economy would benefit from cabotage liberalization, see Warner, *supra* note *Ibid.* at 317-318, and note 276.

[79] Dempsey 'Airlines in Turbulence', *supra* note 35 at 91.

[80] Cabotage in the US would be useful for foreign carriers only to extend some of their existing US services to major inland cities. A Seoul-to-Los Angeles-to-Dallas route, for instance, would allow Korean Air Lines to offer single-airline convenience for travelers between Los Angeles and Dallas.

[81] Bliss, *supra* note 78 at 399, note 66.

[82] The third package of the EU liberalization (1992), see Part 1, Chap. 1, para. 1, above.

[83] The US airline industry has always been considered as the strongest airline industry in the world. However, the terrorist events occurring in the US on 11 September 2001 have seriously endangered the national industry, and most of the US airlines currently survived

foreign carrier wants to enter the US market, it would find it difficult to compete. Indeed, though US carriers currently live an un-precedent financial crisis, it is still unlikely that liberalization of the US domestic market would create the threat of potential foreign dominance.[84]  In their comparison of US and European carriers, Gibson and Goldstein note that:

> The US carriers have dramatically lower costs, a larger domestic passenger base, greater experience running hub and spoke-networks, and sophisticated management and pricing systems. (...) Since few foreign carriers mount heavy transpacific or transatlantic schedules, the number of foreign carrier flights actually added to US airways would be minor. Many of these would be at off-peak hours, due to the timing of the intercontinental flights they meet or continue. Would Singapore Airlines set up a huge hub in San Francisco and offer deep discount flights throughout the US? I doubt it. Such a foray would be expensive and meet heavy competition. Most of the airline's low-labor cost advantages would disappear operating from a US base.[85]

Considering all the above, the lifting of foreign investment restrictions does not have a detrimental effect on traffic rights, alternative solutions can be found. Nor would it affect national economy since cabotage is not a danger for the national airline industry. Cabotage, fostering national competition by its nature, even benefits the whole industry and the public interest. Consequently, international traffic rights and cabotage do not justify the requirement of the substantial ownership and effective control of airlines.

*2.4.5 Risk of loss of a bargaining chip*  Is the State's bargaining power, resulting from ownership restrictions, essential for the national airlines to expand their route? In a bilateral struggle, a State can effectively expect 'extra' benefits from its partners in exchange for concessions that may be extracted from the other party whose carrier's nationality is in doubt or from a party that owns such a carrier.[86] As van Fenema notes, '[t]he external aspect of ownership and control is thus primarily a matter of airpolitical expediency, not of law or principle.'[87] The issue of bargaining power with respect to ownership and control requirements has played an important role in a number of negotiations, such as in the British Airways/KLM joint venture discussions of 1991 and in the 'Alcazar' negotiations between

---

after the events because of State assistance. The attacks have certainly weakened the US airline industry that will need time to regain its previous position of strength.

[84] Edwards, *supra* note 22 at 628-629.

[85] Warner, *supra* note 5 at 317, note 275.

[86] 'For example, if a foreign airline gains control over a US airline, this foreign airline could essentially obtain complete access to the US aviation market. In these circumstances, the foreign airline's government would have little incentive to grant other US airlines greater access to its own market', see Warner, *ibid.* at 312.

[87] van Fenema 'Consequences and Steps to be Taken', *supra* note 67 at 64.

Swissair, Austrian Airlines, SAS and KLM, thereafter,[88] as well as in KLM/Northwest and BA/USAir transactions.[89]

Despite these several uses, there are increasingly fewer justifications for governments to fear losing their leverage in the negotiating bilateral process in the event of the liberalization of the ownership and control rule. The international airline industry is moving slowly towards a multilateral negotiating process and abandoning bilateralism. Bilateral agreements still remain the main method by which air carriers and the State concerned bargain. However, both the need to lift national constraints (mainly foreign investments and market access restrictions) and the need for equal States to gather their interests within a sole group, lead the airline industry towards a more regional, plurilateral and/or multilateral environment.[90] In this new context, reciprocal relations predominate to the detriment of unequal negotiations that are characterized by bargaining chip power. Some States see this future landscape as unrealistic and too optimistic since there are huge economic differences among the States. The US, especially, believes that reciprocity is not conceivable – it has too much to give and not enough to receive – and its domestic air transport market could be endangered in such a multilateral environment.[91] This argument is in fact no longer true, since the superiority of the US airline industry is less and less obvious due to changing circumstances and new trends such as the current air transport crisis directly affecting the US industry, the alliance phenomenon, airline concentration (more and more in Europe), and the development of regional agreements. Air carriers can increasingly play on an equal basis, which makes reciprocal relations more possible. Hence, national leverage in the negotiating bilateral process becomes less vital for airlines to survive. Furthermore, a possible future shift in the airline industry would make the use of ownership restrictions as a bargaining chip ineffective. If traffic rights are liberalized prior to the ownership and control restrictions, States would no longer pursue the same goals in their bilateral negotiations, namely, to try to get traffic rights from their partners to the greatest possible extent and by any means. Thus, as a first step, the liberalization of traffic rights would allow the airline industry to play on the worldwide scene on a less political basis.

Accordingly, considering the evolutionary environment and the current needs of the airline industry, the lack of State bargaining power would not endanger national carriers. Therefore, the liberalization of ownership and control should proceed, given that States, and especially the strongest ones, do not have to fear the loss of their bargaining power in their possibly more equal future air transport relations.

---

[88] *Ibid.*
[89] See Part 1, Chap. 2, para. 2, above.
[90] See Part 1, Chap. 1, para. 2, above.
[91] Edwards, *supra* note 22 at 620.

## 3. Aviation Safety and National Security Justifications

Given that safety and security are the two most important features of the airline industry,[92] is the substantial ownership and effective control requirement a prerequisite to maintaining effective aviation safety and national security?[93]

While legal and economic needs are flexible, safety and security obligations are not. National measures are required to deal with situations of emergency or threat, therefore, the consequences of carelessness in this regard directly concern the public interest. These two absolute air carrier obligations make the airline industry particularly unique. The International Transport Workers' Federation (ITF) stresses the need to maintain a close link between safety and security requirements and the economic regulatory framework within which such services are provided.[94] In its paper, ITF explains that significant weaknesses in security arose directly as a consequence of the economic deregulation of aviation. The purpose of this section is to expose the alternative solutions to maintaining the highest required level of safety and security and pursue the air transport liberalization at the same time.

### 3.1 Aviation Safety and Flags of Convenience

'Would increasing permissible foreign investment in [national] airlines raise legitimate safety issues? Could foreign ownership of a [national] airline render that airline less responsive to [national authority] safety oversight?'[95]

Aviation safety is related to the state of the aircraft itself. High safety standards are defined by States and by the international community (ICAO) in order to ensure the highest possible level of passenger security.[96] Nevertheless, countries still have different levels of safety standards; while some require a high

---

[92] R.I.R. Abeyratne, *Emergent Commercial Trends and Aviation Safety* (Aldershot: Ashgate, 1999) at 165 [hereinafter Abeyratne *Emergent Commercial Trends*].

[93] The usual distinction in air transport is between aviation safety and aviation security. Safety relates to the prevention of accidental events that can affect material or people (design of aircraft, maintenance, etc.), while security is the prevention of intentional acts which aim to affect planes or people (hijacking, bombs, etc.). In the present study, safety and security will be analyzed in the context of the ownership and control issue. Thus, aviation safety will be studied according to its current definition, whereas security will not be studied regarding aviation security, but regarding national security. National security does not aim to protect the air transport but the State as a whole.

[94] ICAO, Working Paper (*The Orderly Evolution of Air Transport Services: Secure and Safe Economic Regulation in an Area of Globalisation*) No. A33-WP/227 (28 September 2001).

[95] ABA doc., *supra* note 1 at 21.

[96] In 1997, ICAO has launched the Safety Oversight Programme, by which experts have assessed the capacity of participating States to control the level of safety for which they have responsibility. The program has been extended to personnel licensing, operation of aircraft and aircraft airworthiness. The State's audit is almost complete today; it has represented a successful operation approved by the whole community of States, see Abeyratne *Emergent Commercial Trends*, *supra* note 92 at 166.

level of safety before accepting the registration of an aircraft, others remain much less stringent. In the latter case, it is therefore less expensive for airlines to register their fleet in these countries, as they do not have to comply with strict criteria. Under the current regime, an air carrier cannot easily transfer its registry because the vast majority of countries still require that the carrier be substantially owned and effectively controlled by nationals of the State. The main concern is that if this last restriction is lifted, it will become possible for an air carrier to transfer its registration. Hence, the risk of the 'flags of convenience'.[97] This concept has been used in maritime law for decades.[98] Indeed, ship owners can transfer the registration of their vessels to more favorable countries of registry to avoid burdensome taxes, potentially high liability exposure for maritime disasters, and less stringent safety requirements. The US has been very concerned about this problem, because as a result of the high costs of compliance with the US requirement of registration, many American ship owners have decided to 'reflag' their vessels. In 1995, the US realized that the US fleet would vanish entirely unless decisive steps were taken to support the merchant marines. As a result, the *Maritime Security Act* was adopted in 1996,[99] which states that US ships would be subsidized over the next decade; in return, carriers must participate in the Maritime Security Fleet.[100] Despite these efforts, US ships and other national fleets continue to decline, departing and registering in 'open registry' countries,[101] which makes their safety more dubious.

Regarding this situation, should the airline industry be concerned about a similar risk of the flags of convenience and lax safety standards in the event of cross-border investment developments? Several reasons have been raised so far to demonstrate that aviation safety would not be endangered by a change to the ownership and control principle.

First, the aviation and the maritime sectors are quite different on many points, therefore, situations faced by both sectors are not necessarily analogous. One of the differences is the meaning and application of the ownership and control requirement with respect to each of the two fields. Due to the bilateral regime, the requirement in the aviation sector prevents or at least impedes the cross-border mergers of air carriers. By contrast, the sale of US shipping companies to foreign shipping lines is not as affected by statutory ownership requirements. While the US DOT closely examines the links of the US owners of US air carriers with foreign

---

[97] 'A flag of convenience is used to describe the flags of such countries...whose laws allow – and indeed make it easy – for ships owned by foreign nationals or companies to fly the flags. This in contrast to the practice in the maritime countries (and in many others) where the right to fly the national flag is subject to stringent conditions and involves far reaching obligations', see Kass, *supra* note 48 at 150, note 44.

[98] The American Bar Association raises the following issue: 'To what extent are the ownership and control requirements in the maritime and defense industry instructive?', see ABA doc., *supra* note 1 at 20.

[99] *Maritime Security Act*, Pub. L. No. 104-239, 110 Stat. 3118 (1996).

[100] For further discussion on US maritime laws, see Bohmann, *supra* note 21 at 730-738.

[101] 'Open registry' countries are the countries with less strict registration requirements that permit the registration of almost all ships owned by foreign nationals.

interests, the US Maritime administration (Marad) does not seem to be equally interested in the ultimate ownership situation of US shipping companies that receive federal subsidies. As Bohmann states, 'it appears as if Marad views the ownership status of a shipping company and potential foreign interests in the company as less important than the continuing maintenance of a US-flagged and US-crewed shipping line.'[102] In addition, the aviation and the maritime sectors have a very different legal regime. Restrictions on traffic rights do not exist in the maritime sector, whereas in aviation, an airline can only exploit traffic rights that are designated to that carrier by the carrier's State of registry. Thus, an airline will always choose a country of registry that owns attractive and profitable traffic rights, rather than a country with low costs and safety standards.[103] Accordingly, the risk of the flag of convenience is unlikely to have a great influence on the aviation sector.

A second reason that explains why the international community should not fear an extreme diminishment of safety standards is the existence of an international framework that defines uniform standards in certification that ensure civil aviation safety. As a result, the question of aviation safety should not be linked to and depend upon the foreign ownership and control issue.[104] 'Ownership' and 'safety' are two different issues that must and can be resolved separately. Thus, as a primary concern, the international community of States is bound to certain safety standards defined by the *Chicago Convention* and should take into account international Standards And Recommended Practices (SARPs) drafted by ICAO. Provisions of the *Chicago Convention* dealing with safety (Articles 11, 12, 31, 32 and 32(b)), impose strict rules on States to ensure high safety standards for international civil aviation.[105] States are bound to take ICAO SARPs into account (Articles 37 and 38 of the *Chicago Convention*), in order to maintain a uniform safety framework.[106] Complementing this regulatory framework on safety, ICAO established the Universal Safety Oversight Audit Programme (USOAP) in January 1999.[107] Accordingly, obviously, safety does not depend on national foreign investment policy, but rather on whether States comply with the minimum safety standards.

A third reason also demonstrates how aviation safety would not be jeopardized by any change in the ownership and control principle. ICAO proposed an alternative solution in 1998 in order to liberalize the ownership and control

---

[102] Bohmann, *supra* note 21 at 737.

[103] *Ibid.* at 727.

[104] 'While governmental concern for ensuring safety is laudable, it is not an objective that is necessarily related to foreign ownership' notes Kass, *supra* note 48 at 150-151.

[105] The *Chicago Convention*, *supra* note 10. For further comments on these Articles, see Abeyratne *Emergent Commercial Trends*, *supra* note 92 at 167-168.

[106] SARPs are enounced in 18 Annexes of the *Chicago Convention*, see Abeyratne *Emergent Commercial Trends*, *ibid.*; M. Milde, 'Enforcement of Aviation Safety Standards' (1996) 45 Abhandlungen 3 at 4-9. The program has been very well accepted by States, e.g. by the US, see C. Shifrin, 'FAA plans safety change' *Airline Bus.* (June 1999) 11.

[107] See Part 2, Chap. 3, para. 3, above; for the safety program outcome, see ICAO, *Annual Report of the Council*, ICAO Doc. 9770 (2001) 10 [hereinafter ICAO Annual Report 2001].

principle without weakening aviation safety. The European Civil Aviation Conference (ECAC) approved the solution and addressed it as follows:

> In recommending the liberalization of the ownership and control provisions, the ICAO Air Transport Regulation Panel stated that provision should be made for a strong link to remain between the airline and the designating State, primarily to ensure that there would be no confusion about which State was responsible for a carrier's safety and to prevent the emergence of 'flags of convenience' airlines. A strong link can be ensured by requiring both the following elements: that a carrier's principal place of business be in the country which designates it; and that a carrier hold an Air Operator's Certificate from the country designating it.[108]

Accordingly, broadening the criteria of substantial ownership and effective control by including the 'strong link' requirement in the bilateral agreements, seems to be a good 'compromise' between full liberalization and the priority for safety. Indeed, the 'strong link' criterion is based on the concept that an airline operation will essentially remain an enterprise that has a single identifiable geographic base. Therefore, it can prevent airlines from registering in unreliable States with less stringent or unknown safety rules.

In addition, in 1998 the ECAC proposed a model standard bilateral clause on safety[109] to include in the bilaterals, which would enable States to withdraw a foreign airline's permit if there were grounds to suspect that the safety of its operations fell short of international standards. By this clause, safety would be better ensured, and would be resolved independently of the ownership and control issue. Therefore, whatever States decide about the liberalization of ownership, aviation safety would not be endangered.

---

[108] ECAC, *Task Force on Ownership and Control Issues, First Meeting*, ECAC Doc. OWNCO/1 – WP/2 (24 December 1998) [hereinafter ECAC Doc. 1998]. To define the 'strong link', ICAO recommends that: '[i]n judging the existence of a strong link, States should take into account elements such as the designated air carrier establishing itself, and having a substantial amount of its operations and capital investment in physical facilities in the designated State, paying income tax and registering its aircraft there, and employing a significant number of nationals in managerial, technical and operational positions', see ICAO, *Policy and Guidance Material on the Economic Regulation of International Air Transport*, ICAO Doc. 9587 (1999) at 2.2.

[109] ECAC Doc. 1998, *ibid.*, ECAC recognizes that 'one of the elements of the ECAC/JAA aviation safety action programme comprises the strengthening of controls over the safety of foreign aircraft (...), that provision in bilateral agreements could provide a basis for strengthening controls over safety of foreign aircraft through both general provision and provision enabling random ramp checks, that is desirable for Member States to have available to them a safety model aviation safety clause for incorporation into their bilateral agreements.'

*3.2 National Security*

How could national security objectives be affected by changes in foreign investment rules in [national] airlines? Would it be possible to achieve these objectives by other methods if the rules were changed?[110]

National security is the most specific feature that makes the aviation industry so unique, compared to any other industry, and so important for States. Indeed, States need to maintain a strong domestic airline industry capable of serving loyally the nation in times of crisis, and the security can be better ensured if the industry is owned by nationals. National security being the main priority of the US government, the US has always been wary of liberalization of ownership and control, and for the same reason the DOT has scrutinized airline alliances (e.g. the KLM/Northwest alliance).[111] The exercise of US security is implemented by the Civil Reserve Air Fleet (CRAF) program, established in 1952 under a *Memorandum of Understanding between the US Departments of Commerce and Defense*.[112] This program was executed pursuant to Executive Order 10219, wherein President Truman directed the Secretary of Commerce to formulate plans and programs for 'the transfer or assignment of aircraft from civil carriers to the Department of Defense (DOD), when required to meet needs of the armed forces.'[113] The CRAF program was activated for the first time in 1990 during the Gulf War, and the successful operations proved the real value of the program.[114] Preserving US national security remains a top priority today.

Will national security still be guaranteed for future emergency missions if the level of foreign investment in US carriers increases? Are restrictions on ownership and control a prerequisite to preserving national security? The debate on this point has always been, and still is, very dynamic. National security may be the only reason that truly justifies foreign investment restrictions. Apart from a few authors who believe that the national security argument against foreign ownership is just an excuse used to protect the national economy,[115] the debate focuses on the possibility to ensure national security protection in the absence of restrictions on ownership and control. Several alternatives are listed.

First, the government could ensure that aircraft remain available for national defense purposes, by making participation in the CRAF a condition for registering

---

[110] ABA doc., *supra* note 1 at 21.

[111] Arlington, *supra* note 22 at 162-164.

[112] C.M. Petras, 'Foreign Ownership of US Airlines and the Civil Reserve Air Fleet Program: Cause for Concern?' 15 March 2001, at 3, note 6 [unpublished].

[113] *Ibid.* at 3, note 7.

[114] For more details about the CRAF operation during the Gulf War, see *ibid.* at 5; see Arlington, *supra* note 22 at 161-162; Edwards, *supra* note 22 at 640.

[115] '[I]nternational airlines, in many countries including the United States, have sought to stabilize their positions in an increasingly competitive global marketplace by playing the trump cards of national security and prestige but keeping hidden their real agenda of protectionism', see R.D. Lehner, 'Protectionism, Prestige, and National Security: The Alliance Against Multilateral Trade in International Air Transport' (1995) 45 Duke L.J. 436 at 450.

an aircraft in the United States.[116] This alternative is not adapted at all to a war situation because 'even if foreign investors consented to participate in CRAF, the issues relating to the airline's willingness to serve the nation in times of crisis would have a deleterious effect on reliability of the CRAF fleet.'[117] This problem occurred during the Gulf War, when some foreign ships refused to make deliveries to certain ports,[118] severely limiting their usefulness.[119] Furthermore, considering the political instability of certain countries today, the US do not know which countries they may trust.[120] Since international airlines may act as instruments of their governments, the US cannot afford to take the risk of losing the fleet of a US airline, which is owned by an unreliable foreign airline in times of emergency and war.[121]

The possibility to commandeer aircraft is a second alternative. In other words, if a foreign-owned airline violates its CRAF agreement and refuses to provide its aircraft in times of crisis, the DOD could simply commandeer the aircraft it needs. In reality, though, this option could create political problems with the State(s) concerned. By doing so, the US would risk increasing international tensions: 'it would clearly be a less than desirable *modus operandi* for the CRAF program.'[122]

The purchase of aircraft by the DOD is a third solution. It is problematic for two reasons. In case of emergency, there no time to create an entire fleet: looking for new aircraft, flight and ground crew is time-consuming and would render the CRAF program inefficient. Moreover, purchasing aircraft would be an inadvisable financial operation for the US government since it would generate tremendous costs in terms of operation, maintenance, and personnel, it would consequently remove all of the financial benefits of the CRAF program.[123]

---

[116] Petras, *supra* note 112 at 8.

[117] *Ibid.*

[118] The same program as the CRAF program exists in the maritime sector. It is called the Maritime Security Program (MSP) and it provides subsidies to US carriers that make their vessels available for military use during a national emergency, see Bohmann, *supra* note 21 at 736-737 and note 192; on the concern to preserve US citizenship requirements for MSP vessels operators, see 'US questions Maersk's power' *FairPlay* (19 July 2001) 22.

[119] Edwards, *supra* note 22 at 641.

[120] 'While commercial aviation is becoming increasingly global, political alliances between nations are constantly shifting. Iran, which was once the United States' closest ally in the Arab World, is now a sworn enemy, while Russia, which was dubbed "the Evil Empire" and "the force of evil in the world" during the 1980s, is now an important international partner of the United States' and this context is in constant evolution, see Petras, *supra* note 112 at 11.

[121] 'During times of crisis we need to know without question that there is support', see K. Walker, 'US DoD gives Red Light to Ownership Changes' *Airline Bus.* (June 1999) 11.

[122] Petras, *supra* note 112 at 10.

[123] 'There is no prospect that the US Congress will provide the DOD with the financial means to acquire a significantly larger organic fleet of transport aircraft', see IATA, Government and Industry Affairs Department, *Report of the Ownership & Control Think Tank World Aviation Regulatory Monitor*, IATA doc. prepared by H.P. van Fenema (7

Despite the strength of the national security argument against the abolition of foreign investment restrictions, the US DOD should reconsider the question. In case of unlimited foreign investments, the US government fears the doubtful loyalty of foreign airlines and the lack of patriotism that would be required to ensure security in the US. However, although this notion of patriotism is an important stimulus, it is not the primary motivation of CRAF participants. Economic benefit is the main incentive.[124] Thus, such an incentive should function in a structured multilateral framework, and foreign airlines, as well as American airlines, would likely do the job of contributing to the security of the United States. Furthermore, additional legal alternatives can secure the defense commitment of the US. Specific contracts could be drafted that would safeguard the necessary supply of aircraft. The combination of a contractual commitment with a personal liability of directors and managing officers would raise the likelihood of compliance with the CRAF program.[125] Moreover, even if a foreign airline were to obtain control over a US air carrier, the effects of this transaction on US national security would, if necessary, be reviewable by the President of the United States under Section 101(16) of the *Federal Aviation Act*. When he considers appropriate, the President can suspend or prohibit any international acquisition, merger, or takeover in case foreign control threatens to impair the national security.[126] This provision is a possible solution, as the aircraft, in such a case, would probably remain in the US territory. Indeed, foreign carriers would likely not invest in US carriers in order to get access to aircraft and to use their aircraft in their homeland markets.

With a regime of unlimited foreign investment, the US government can even anticipate any inconvenience. For instance, the DOT, in conjunction with other government agencies, could retain the right to disapprove of foreign investment in US carriers if national security is threatened.[127] It can even be legally

---

September 2000) at 60 (Separate comments of Jeffrey N. Shane on 'Airlines and National Security in the US') [hereinafter IATA doc.].

[124] 'While participation in the CRAF Program is optional for US airlines, such participation is a prerequisite to eligibility for doing significant business with the military (49 U.S.C. § 41106)', see *ibid.* at 61.

[125] Bohmann, *supra* note 21 at 712-713; another related idea has been addressed: 'to avoid any doubt about the impact of liberalization on CRAF participation, more specific legal requirements for US-controlled and foreign-controlled carriers alike – possibly implemented in the form of permit conditions – should be seriously considered', see IATA doc. (Separate comments of Jeffrey N. Shane on 'Airlines and National Security in the US'), *supra* note 123 at 62.

[126] Warner, *supra* note 5 at 310 note 238 (50 U.S.C.A. app. § 2170 (West Supp. 1993), 'The President may direct the Attorney General to seek appropriate relief, including divestment relief, in the district courts of the United-States in order to implement and enforce this section'. Section 2170(f)(3) further provides '[f]or purposes of this section, the President [...] may, taking into account the requirements of national security, consider among other factors the control of domestic industries and commercial activity by foreign citizens as it affects the capability and capacity of the US to meet the requirements of national security'.

[127] Kass, *supra* note 48 at 179.

foreseen that, if not satisfied with a foreign carrier's contractual commitment to participate in CRAF, the DOD could require a lower foreign investment limit, as an exception, for foreign airlines willing to participate in the CRAF program, such as the current 25 per cent limit. This last proposal would allow the international airline industry to conclude financial operations, without any constraints, like any other important industry.

## 4. Conclusion

To conclude this third chapter, it has to be emphasized that justifications of the ownership and control regime have obviously evolved since 1946. National security concerns have historically been at the core of the argument against foreign investment in US air carriers. Then, economic reasons were addressed by governments in order to justify foreign investment restrictions. In reality, not many arguments survive scrutiny. Either they are just used as 'excuses' to hide national protectionism and, therefore, do not make sense (e.g. job losses, competition, economic security concerns), or they represent completely different problems that may simply be resolved independently of the ownership and control issue (e.g. the liability regime, cabotage and safety concerns). Regarding the legal and economic reasons, the airline industry does not differ at all from other economic sectors, except that it is not a highly profitable industry. Accordingly, the need of foreign capital is certainly the most relevant issue and therefore represents the main reason in favor of the liberalization of foreign investment. National security is the tricky part of the study, since it may require protectionist measures. National security makes the aviation industry unique with respect to the particular protections required, especially as compared to other industries. More importantly, the public interest is directly served by the proper implementation of the national security program. National security is still a justifiable concern, especially, but not only, for the US, so it might not be the time for legal changes that could threaten national security. Yet the airline industry must move on and evolve, as any industry, towards liberalization and globalization, and alternatives must also be considered. National security seems to be the only credible argument in favor of national restrictions on ownership and control. In this case, it is certainly excessive to use it as the sole impediment to industry growth, despite its great importance. National security is not an insurmountable impediment and solutions may be available, as discussed above. There is no doubt that it would be much more in the public interest to foster foreign investments among international airlines, rather than maintaining unnecessary restrictions. Accordingly, the substantial ownership and effective control requirement should be revoked.

# Chapter 4

# Consequences of the Abolition of the Ownership and Control Restrictions

## 1. Introduction

Since the health of the airline industry requires a liberal regime with respect to foreign investments, and since there are no legitimate reasons that justify the present restrictive regime, the substantial ownership and effective control principle should be reconsidered. What would be the consequences of an overhaul of the regime? This chapter will address these consequences by, first, studying the legal effects of a liberal regime. It will then focus on the economic benefits for the airline industry of the free movement of capital worldwide. As a third point, the following important issue will be raised: what will be the prospective scenario regarding airline consolidation (e.g. mergers) in the event of total liberalization of the ownership and control rule in the international airline industry?

## 2. Legal Effects

### 2.1 The Right of Establishment

The right of establishment is defined as the right of a foreign investor or foreign airline to set up an airline in a given country.[1] This right has been recognized inside the EU in the aviation sector, by the third package of the EU liberalization.[2] Presently, EU nationals have the right to own and operate a carrier in each Member State.

      The right of establishment is an indirect legal consequence of this proposed overhaul of the ownership regime, as it is in fact directly related to cabotage. Thus, in case of removal of ownership and control restrictions, a foreign carrier is entitled to establish, in a country, a subsidiary that operates domestic services, as well as international services from/to that country. Indirectly then, the right of establishment and cabotage are granted to the foreign carrier, given that the subsidiary, owned by a foreign airline, is considered, in that respect, as a national company. In fact, even though the ownership regime overhaul makes the cabotage

---

[1] P.P.C. Haanappel, 'Airline Ownership and Control and Some Related Matters' (2001) 26-2 Air & Space L. 90 at 97 [hereinafter Haanappel 'Airline Ownership and Control'].
[2] EU, *Council Regulation 2407/92 on Licensing of Air Carriers*, [1992] O.J. L. 240/1 [hereinafter *Council Regulation 2407/92*]; see also Part 1, Chap. 1, para. 1, above.

restrictions non-applicable, the right of establishment would be established. This statement can in fact support the necessity of revamping the ownership regime because, as discussed above, the foreign carrier would invest in a national business, this business would be a national operator and, therefore, national law would apply. Consequently, a liberal regime on ownership and control does not have a negative impact on the national system. For instance, if a European carrier operates within the US market, it will be subject to the full application of US laws, including tax and regulatory measures, just like US nationals are. Its employees would also be taxed and regulated at the federal and State level:

> In other words, any airline that participates in US domestic commerce – whether US or foreign owned – will have to bear all the burdens, and not just reap the benefits, of US Commerce. (...) [I]t is neither realistic nor equitable as a matter of law or policy to allow foreign air carriers to participate as foreign carriers in US domestic air services. The proper means of opening US markets to those carriers governed by foreign law is to confer upon them a right of establishment, i.e., the right to own and operate a carrier that complies with US laws – in other words – a US air carrier[3]

In sum, foreign nationals that participate in any national domestic air commerce will have to participate on the terms that would be generally applicable to all companies that participate in other domestic sectors of the economy. Thus, the right of establishment seems to be a logical continuity of the airline liberalization process: it would allow the air industry to evolve like any other main industry since no different legal treatment would be given to all companies established in a given country.

## 2.2 *Extraterritoriality*

With the growing globalization of air transport, airlines, particularly in the US and in Europe, are increasingly eager to form alliances with foreign counterparts in order to enhance the efficiency, scale and scope, of their international operations. In addition, a revamping of the ownership regime would lead to concentration through mergers and takeovers. All these developments inevitably raise antitrust issues, especially considering that the aviation industry is already dominated by a relatively small number of market participants.[4] More importantly, these operations create legal conflicts between the national laws. Hence, the increasing importance

---

[3] IATA, Government and Industry Affairs Department, *Report of the Ownership & Control Think Tank World Aviation Regulatory Monitor*, IATA doc. prepared by H.P. van Fenema (7 September 2000) at 58-59 (Separate comments of Warren Dean on 'The Right of Establishment') [hereinafter IATA doc.].

[4] G.P. Elliott, 'Antitrust at 35,000 Feet: the Extraterritorial Application of United States and European Community Competition Law in the Air Transport Sector' (1997-1998) 31 Geo. Wash. J. Int'l L. & Econ. 185 at 187.

of extraterritoriality, given that '[e]xtraterritoriality is one concept which could affect more than one jurisdiction in the application of domestic trade law.'[5]

In the previous paragraph, it was observed that national law is applicable to any company operating a domestic service, whether the company is in national or foreign hands, since the domestic market is affected. This means that the applicability of law is not linked to ownership, but to national interests. By the same logic, international airline mergers, affecting several domestic markets, will have to comply with all the national legal regimes involved and, more specifically, with national competition laws (antitrust laws).[6] This concept of extraterritoriality of national law has been recognized by both the US courts and the Court of Justice of the European Communities. The Court of Justice has justified the application of national laws when an activity has an effect on the domestic market and is implemented within the European Community, even though the commercial activity is outside the territory, the activity is conducted by foreign citizens, or the company is owned by foreigners.[7] While the US have always had a wide scope of extraterritorial jurisdiction in respect of anti-competitive practices, the EU has become more and more active in recent years in this respect, as it has disapproved of mergers of US companies in the aviation industry. Two cases in the aircraft manufacturing industry can be taken as examples: the European Commission objected to two major mergers in the US, which had already been approved by the US authorities, lest they lead the companies to a dominant position on the international market. The first one was the Boeing/McDonnell Douglas merger in 1999;[8] the other objection concerned the 2001 General Electric Co. proposal to take over its rival Honeywell.[9]

---

[5] R.I.R. Abeyratne, 'Would Competition in Commercial Aviation ever fit into the WTO?' (1996) 61 J. Air L.& Com. 793 at 849 [hereinafter Abeyratne 'Competition in Commercial Aviation'].

[6] 'One cannot deny that in the era of global economy, some degree of extraterritoriality in the enforcement of national competition rules is inevitable', see *ibid.*

[7] For more details about the cases in the US and in the EU, see *ibid.* at 849-852; J.W. Young, 'Globalism Versus Extraterritoriality. Consensus Versus Unilateralism : Is There a Common Ground? A US perspective' (1999) 24 Air & Space L. 209 at 211-213.

[8] 'The EC finally approved the merger of the companies on 30 July 1997, but only after Boeing had acceded to major concessions, resulting partly from political pressure from the Clinton Administration', see K. Luz, 'The Boeing-McDonnel Douglas Merger: Competition Law, Parochialism, and the Need for a Globalized Antitrust System' (1999) 32 Geo. Wash. J. Int'l L.& Econ. 155 at 155; T. O'Toole, 'The Long Arm of the Law' – European Merger Regulation and its Application to the Merger of Boeing & McDonnell Douglas' (1998) 11 Transnat'l L. 203 at 203; P.S. Dempsey, 'Competition in the Air: European Union Regulation of Commercial Aviation' (2001) 66 J. Air L. & Com. 979 at 1117-1122 [hereinafter Dempsey 'Competition in the Air'].

[9] 'In its appeal against the Commission's July [2001] ruling blocking the world's largest industrial merger, GE is understood to have launched a stinging attack on the theories and practices that are used by the European antitrust authorities. (...) The Commission said that the combined group would have driven rivals out of the market by offering a combination of engines and avionics products' see F. Guerrera, 'GE fires salvo at European Commission' *Financial Post* (5 November 2001) FP11; P. Bocev, 'l'Europe recale le mariage GE-

These cases also highlighted the need to harmonize the rules of competition in the international air industry, especially between the US and the EU antitrust laws.[10] Globalization requires more cooperation and should lead to concentration; therefore, if the airline industry begins to concentrate internationally as a result of revisions of the ownership regime, a uniform, or at least harmonized regime of antitrust laws[11] will need to be agreed upon by the international community in order to avoid any discrepancies and uncertainties which create impediments to a healthy development industry.[12]

## 3. Economic Benefits

The liberalization of the ownership and control rules will enable increased foreign direct investment (FDI) in airlines. This will certainly benefit the entire aviation industry and the public interest. In his book on International Business, Charles Hill describes the costs and the benefits of FDI to countries, not specifically with regard to the aviation industry, but in general. He asserts that States tend to adopt a pragmatic stance, pursuing FDI policies to maximize the national benefits and minimize the national costs of FDI.[13] In fact, most of the benefits outweigh the costs of foreign investments in a national industry, whether with regard to an airline or any other company. Some important benefits will be addressed, using the following example: a European airline (A) takes a more than 50 per cent share in a South-American airline (B). To use the terms of Hill, A is considered as the airline of the 'home-country' / the country who invests / and B the airline of the 'host-country' - the country who receives the investment.

An increase of foreign investments would benefit the South-American industry. By creating far-reaching cooperation, a foreign company will be able to make 'a positive contribution to the host economy by supplying capital, technology, and management resources that would otherwise not be available and thus boost that country's economic growth.'[14] These resources benefit the airline industry. In case the airline B is in need of capital to expand, or just to survive and

---

Honeywell' *Le Figaro* (4 July 2001) 1; A. Chuter, 'Growing pains: Differing approaches taken by US and European regulators over the proposed GE/Honeywell merger highlight a need for common guidelines' *Flight Int'l* (19 June 2001).

[10] 'If ever there was an illustration that the United States of America and the United States of Europe are separated by more than water, it is General Electric's efforts on both sides of the Atlantic to gain approval for its merger with Honeywell', see Chuter, *ibid.* at 5.

[11] For a complete study of the competition regimes in the EU and in the US, with respect to international airline alliances, see A.C., Lu, International Airline Alliances : EC Competition Law/US Antitrust Law and International Air Transport (The Hague: Kluwer Law International, 2003), see also Elliott, *supra* note 4 at 196-204.

[12] This issue of the harmonization of competition laws will be addressed in Part 3, Chap. 5, para. 5, below.

[13] C.W.L. Hill, *International Business: Competing in the Global Marketplace*, third ed. (Boston: Irwin/McGraw-Hill, 2001) at 211 [hereinafter Hill 2001].

[14] *Ibid.* at 213.

avoid bankruptcy, a contribution in technology would allow B to enhance its productivity, as technology can be incorporated in a production process (e.g. aircraft maintenance) or in a product (e.g. personal computers). Furthermore, foreign management skills, acquired through FDI, may also produce important benefits for B, such as the training of B's personnel to occupy managerial, financial, and technical posts in the new enlarged company. In fact, these three spin-off effects concern airlines from both developed countries[15] and developing countries.[16] Indeed, the absence of national restrictions on ownership and an increase of foreign capital input would enable weaker States, with limited sources of capital, to keep a carrier inside their territory. In addition, allowing the free movement of capital in the airline industry could enhance employment in the company B because of its expanded activity, a direct consequence of A's capital input. Employment could increase in B's country as well, due to the positive effects of B's growth on all related activities, e.g. on airports and tourism-related enterprises, and, of course, as well as, of increased local spending by B's employees. Through the ensuring increase in consumer choice (e.g. B is able to add services), foreign investment can produce increased competition in B's national market, thereby driving down prices and increasing the economic welfare of consumers.[17] All these possible benefits demonstrate that the international community of States should open the airline industry to foreign investment as it would be profitable for the airlines concerned, as well as for the respective national economies.

FDI would also benefit A and the home-country itself. And the benefits of A would not limit B's profits, in a sense that if the host-country as well as the home-country can take advantage of international financial operations, the entire airline industry will be enriched. Of course, the whole industry would benefit equally from the increase of FDI only if all States liberalize their national regimes with respect to ownership and control.[18] Some fear that foreign ownership of national airlines reduces national employment because personnel costs are lower in certain countries. Thus, in our example, A might hire ground and flight crew in the host-country. Although it is a possible risk, it remains very unlikely in many cases, as airline partners are not always equal. First, airlines care about their 'image' – indeed, what about the national brand image if Air France hires South-American

---

[15] P. Sparaco, 'European Deregulation Still Lacks Substance', *Aviation Wk & Space Tech.* (9 November 1998) 53; as recent examples, Belgium and Switzerland lost their national airline.

[16] For instance, in 1999, Venezuela was without a flag-carrier after Viasa's demise, see D. Knibb, 'No Flag in its Future' *Airline Bus.* (May 1999) 72.

[17] 'FDI's impact on competition in domestic markets may be particularly important in the case of services, such as [air transport] and telecommunications (...), where exporting is often not an option because the service has to be produced where it is delivered', see Hill 2001, *supra* note 13 at 218.

[18] Indeed, if not every State liberalizes on the same basis, some will benefit from their investments outside their territory without letting foreign airlines invest in their domestic market.

crews.[19] Second, the benefits for A would outweigh the costs (risk of jobs losses) anyway, given that its investments abroad would allow A to expand its network, and therefore its activity. By investing in B, A will penetrate a new market, taking advantage indirectly from traffic rights.[20] A can even get some slots in the host-country, which would provide it with an additional asset and a strategic position on the 'A' market.[21] Render the current ownership and control regime, it is already possible to get some of these benefits through certain specific agreements such as code-sharing or through an alliance, but only to a limited extent. Abolishing foreign investment restrictions would allow national airlines, as investors, to maximize their profits on a worldwide scale.

## 4. Airline Consolidation Limits

By advocating the abolishment of the ownership and control principle, the present study aims to meet the need for the airline industry to adapt to the current economic trends. These trends include regionalism and multilateralism, liberalization and globalization, consolidation and concentration. Where concentration between airlines of different nationalities is concerned, airlines limit themselves to international alliances rather than creating new, multinational airlines. Once foreign investment restrictions are liberalized, mergers and takeovers will ideally prevail. So far, alliances have been a very effective means to make the industry more efficient, given that alliances contribute for 70 per cent of the airline synergies in terms of services, marketing, costs and network.[22] However, the industry increasingly needs to go further in the consolidation process. Mergers will bring the airline industry the benefits applicable to any other industry, such as bargaining power, cost rationalization, skill sharing within the combined company, in addition to the benefits that are specific to the airline business, such as network optimization (increase of slots[23] and market access), and increased market power (by the power of combined schedules, FFP, agency agreements, and corporate contracts).[24]

While international and national entities will fight for a free trade environment in an airline industry without any regulatory constraints, to allow the

---

[19] See Part 2, Chap. 3, para. 2, above.

[20] For instance, this is the way Swissair extended its operations on the European market, before a EU/Switzerland Agreement be signed.

[21] British Airways adopted this strategy when it purchased almost 20 per cent of the capital of the south-African airline Comair.

[22] J. Naveau, 'Les Alliances entre Compagnies Aériennes. Aspects Juridiques et Conséquences sur l'Organisation du Secteur' (1999) 49 ITA Etudes & Doc. at 23-27.

[23] 'Under the current structure, national carriers have a competitive advantage since they own all the attractive slots and have superior access to airport facilities', see R. Polley, 'Defense Strategies of National Carriers' (2000) 23 Fordham Int'l L.J. 170 at 179-180.

[24] 'Careful adherence to the "10 commandments" of airline post-merger management will ensure airlines get the most out of a merger', see McKinsey & Company, 'Making Mergers Work' *Airline Bus.* (June 2001) 110 at 110-114.

industry to continue growing, the interest of this fight can be questioned. Indeed, in the long-run, will the free movement of capital among international airlines lead to a highly consolidated industry, similar to what is occurring in other main international industries, or, on the contrary, will the industry consolidation and growth still be limited? Three limits seem to confine the airline industry's growth to a great extent, whether foreign investments are restricted or not.

The first limit is competition law, a regulatory constraint that is applicable to all other industries. Merger control policies monitor national and international airline concentration and tend to limit cooperation among airlines in order to prevent the formation of monopolies (US anti-trust regulations) or in order to prevent an undertaking from achieving a market position that makes competition impossible or that injures the consumer (EU merger regulation).[25] Thus, in EU, the merger regulation[26] applies to all concentrations (i.e., where two undertakings merge or where an undertaking acquires part or whole ownership of another undertaking, which have a 'Community dimension').[27] The Commission examines disparate factors regarding the airline concentrations,[28] and if it 'determines that a concentration does strengthen an undertaking's dominant position, or gives rise to one, the regulation requires the Commission to issue a decision declaring the concentration incompatible with the common market.'[29] This regulation has been applied a number of times since its introduction as a result of the increasing number of cross-border airline acquisitions within the EU.[30] As for US anti-trust regulation, the essential substantive provisions can be found in the Sherman Act, Sections one and two.[31] Section one declares every contract illegal that is 'in restraint of trade or commerce among the several States, or with foreign nations'; Section two declares a person guilty of a felony 'every person who shall monopolize (...) any part of the trade or commerce among the several States, or with foreign nations.' Congress passed Section two to ensure that no single person or persons engage in the willful acquisition or willful maintenance of power 'to

[25] Dempsey 'Competition in the Air', *supra* note 8 at 1102.

[26] 'Competitiveness in the EU's air transportation sector is governed by two principal mechanisms, the competition rules (*EU, Council Regulation 3975/87* and *EU, Council Regulation 3976/87*, see Part 1, Chap.1, para. 1, above, which arise from Articles 85 and 86 of the Treaty of Rome), and the merger regulation (*EU, Council Regulation 4064/89 on the Control of Concentrations Between Undertakings*, [1989] O.J. L. 395/1), which is derived from Articles 87 and 235 of the *Treaty of Rome*', see *ibid.* at 1102; for an overview of the merger regulation and its application before 1998, see L. Gorton, 'Air Transport and EC Competition Law' (1998) 21 Fordham Int'l L.J. 602 at 615-619.

[27] *EU, Council Regulation 4064/89*, Article 3(1) and Article 1(1), *ibid.*

[28] The factors are the necessity of preserving or developing effective competition within the Union, the economic and financial power of the undertakings, the interests of all concerned consumers, etc., see Dempsey 'Competition in the Air', *supra* note 8 at 1114; see also J. Balfour, 'Airline Mergers and Marketing Alliances – Legal Constraints' (1995) 20 Air & Space L. 112 at 114.

[29] Dempsey 'Competition in the Air', *ibid.* at 1115.

[30] Polley, *supra* note 23 at 187.

[31] *Sherman Antitrust Act*, 15 U.S.C. §§ 1-2 (1994).

foreclose competition or gain a competitive advantage, or to destroy a competitor.'[32] This regulation applied, for instance, in 2002 to the United Airlines' plan to purchase US Airways.[33] In Canada, competition policy is similar to the one that exists in the US. The *Competition Act* regulates mergers at Articles 91 through 100, stating in its Article 92:

> Where, on application by the Commissioner, the Tribunal finds that a merger or proposed merger prevents or lessens (...) competition substantially in a trade, industry or profession ... the Tribunal may, subject to sections 94 to 96, in the case of a completed merger, order any party to the merger or any other person to dissolve the merger (...) .[34]

These merger and antitrust regulations will reduce the impact on the airline industry, produced by the mergers and acquisitions made possible by the abolishment of the ownership and control principle. Indeed, national competition authorities tend to favor more competition both within their countries, and increasingly on a worldwide scale. This goal is served by the extraterritorial application of their competition laws to prevent the respective market from becoming too highly concentrated. As discussed above, market consolidation has been blocked a few times by the European competition authorities, lest the companies acquire a dominant position in the international market concerned. The first case concerned the Boeing/McDonnell Douglas merger in 1999; the other was the 2001 General Electric Co. proposal to takeover its rival Honeywell.[35]

Two other aspects, specific to the airline industry, limit the effects of foreign investment liberalization. First, given that the airline industry is not a highly profitable one by nature, as a result of the very high operational costs of the air carriers, the number of potential investors, whether they are financial institutions or airlines, in the air transport industry is much lower than in other industries. Therefore, even in an environment of free movement of capital, investors may have little interest in purchasing stakes in foreign airlines. Nevertheless, a difference can be noted among the investors; financial investors would probably be more eager than airline investors to invest in air carriers, as they look for dividends and increased value of stocks, whereas airline investors look for different things, such as economies of scale and scope. Financial investors may have interests in taking minority stake in an airline, airline investors probably not. Thus, US airlines have shown no interest in investing in European airlines so far. In fact, an increase of foreign investment would surely benefit the air carriers of developing countries – their need of foreign capital is sometimes very important as developing States may not afford feeding their national carrier – but investors must

---

[32] Elliott, *supra* note 4 at 197, note 83.

[33] D. Field, 'Regulatory Hurdles Remain for United's Merger Plans' *Airline Bus.* (April 2001) 13.

[34] *Competition Act*, RS, 1985, c. C-34, s 1; RS, 1985, c. 19 (2nd Supp.), s. 19.

[35] The Boeing-McDonnell Douglas merger in 1999, see Luz; O'Toole; and Dempsey, *supra* note 8; the 2001 General Electric Co. proposal to take-over its rival Honeywell, see Guerrera; Bocev; Chuter, *supra* note 9.

find a reason for investing in these airlines. The third aspect which reduces the effects of the foreign investment liberalization are the logistic and infrastructure constraints specific to air transport. Two main concerns illustrate this material problem: most of the airports in the world are already saturated and overcrowded. In addition, air routes are also not unlimited, and even if they were, the Air Traffic Control (ATC) would not be able to come effectively with the resulting operations.

## 5. Conclusion

This fourth chapter can be concluded by stressing once more the importance of revamping the substantial ownership and effective control principle. A liberalization of foreign investments in the airline industry would benefit the 'home-countries' and the 'host-countries' in their reciprocal economic markets. Legally, an increasing international cooperation will affect the national legal systems, especially with regard to the applicability of the law, which will require legislators to make a particular effort to harmonize the national regimes, in particular in the field of competition. Furthermore, the 'autonomous' limits on airline consolidation reduce to some extent the importance of ownership and control as an impediment to growth. Indeed, even though there is no doubt that the current regime should be revised, the international community should be conscious that this revision will not remove all the constraints that make the airline industry a difficult industry; its fragile financial features and the limited availability of infrastructure continue to hamper the industry's prospects for growth and financial health.

# PART 3

# CROSS-BORDER INVESTMENTS LOOM ON THE HORIZON:

## THE STEPS TO BE TAKEN TOWARDS ACHIEVING LIBERALIZATION

# Introduction to Part 3

Cross-border investments must be totally liberalized in the airline industry. Alternative solutions have been proposed by national and international entities to foster foreign investments without liberalizing ownership restrictions. However, these alternatives would only serve to strengthen the complexity of the ownership and control issue. For instance, there has been the suggestion to separate ownership from control, i.e. to allow substantial foreign investments in national airlines and still retain national control of the industry, in order to meet various domestic, economic and regulatory concerns.[1] This solution will in fact not significantly change the current restrictive landscape since the control criterion is already the main factor that determines a country's attitude to foreign investments. Another proposed alternative solution is setting a minimum international standard of 49 per cent foreign voting equity interest in national airlines with the option for States to raise or eliminate the cap. Even though this alternative would surely benefit the industry, it is still not an acceptable solution since foreign investment restrictions will still prevail. Accordingly, the present analysis advocates the abolishment of all national constraints regarding foreign investment in air carriers, for the international community to reform the regulatory framework necessary to help the air transport industry to evolve and adapt to the global economy. The idea of harmonizing national restrictions, suggested ten years ago, is no longer relevant in the present global environment. At present, most States are moving towards a world without this kind of barriers.

To reiterate, the question is no longer *whether* we should liberalize, but rather *how* we should liberalize. Part three examines the future prospects of cross-border investments. Chapter 5 will concentrate on the *a priori* measures to be taken in order to facilitate the abolition of foreign investment restrictions. Chapter 6 will examine the possibility of eliminating the ownership restrictions at the regional level as a precursor to the elimination of restrictions worldwide. Chapter 7 will analyze to what extent the three relevant international organizations, OECD, WTO, ICAO, are involved in the ownership issue, and address the question of what role they can play in the reform process.

---

[1] See Part 3, Chap. 6, para. 2, below.

Chapter 5

# Necessary Measures Prior to the Removal of Foreign Investment Restrictions

## 1. Introduction

The liberalization process of foreign investments cannot be achieved directly without resolving other concerns that have greater priority. Some concerns are particular to the airline industry, such as the traffic rights issue and the safety and security matter. Other concerns are indirectly related to this industry, such as State behavior and the legal harmonization issue. These various questions should be settled as soon as possible to make the ownership and control liberalization process function effectively.

## 2. Liberalization of Traffic Rights

As we saw earlier,[1] the bilateral exchange of traffic rights has enabled national authorities to tailor market access to the needs of the national airline concerned. By carefully maintaining the bilateral balance of often limited opportunities, national authorities protect the airlines and the national interests which the airline serves. Liberalization, in the sense of a more freely granting market access, implies that the airlines are perceived as needing less government protection. Open-Skies agreements, the most liberal version of the exchange of traffic rights, basically give the national airlines concerned total freedom to succeed or fail in the markets, covered by the agreement, to which they have free access (through the virtually unlimited traffic rights).

In a bilateral setting, the benefits, derived from a (liberal) agreement concluded between State B and A, should not be enjoyed by a foreign airline C', the new owner of airline A', particularly if State B and C do not have a liberal agreement (and the airlines of B still do not have sufficient access to the B-C market).[2] But if all three parties had concluded Open-Skies agreements with one

---

[1] See Part 2, Chap. 3, para. 2, above.
[2] An additional example demonstrates this risk of the loss of traffic rights: 'the proposed KLM-British Airways merger in 2000 would have implied that KLM became British-controlled, since the capitalization of BA was approximately four times that of KLM. The US DOT indicated that if such a merger were to take place, it would deny KLM the routes

another, there would be less inclination on the part of any of the governments concerned, in the absence of any restrictions on the traffic rights for the respective airlines, to revoke or challenge traffic rights in case, in our example, airline A' is taken over by airline C'.[3] Challenging ownership and control of an airline in order to maintain a bilateral balance or avoid airlines from third countries using benefits, becomes a less important option. All airlines concerned have all the traffic rights they need and their governments do not feel the need to protect them through bilateral restrictions imposed on foreign carriers.

If this is correct, then full liberalization through multilateral arrangements, which include all freedoms, including seventh freedom[4] (which one will not yet find in the Open-Skies agreements), will pave the way for the removal of foreign investment restrictions through the abolishment of the ownership and control clauses. Of course, the governments still have a long way to go. But with increasing liberalization, the importance of ownership and control in bilateral agreements, as a means to control the behavior of the airlines concerned, will diminish. In other words, multilateral, global liberalization of traffic rights is, also from that perspective, a worthy goal.

### 3. Harmonization of Aviation Safety and Security Standards

In the aviation safety study in Part 2,[5] it was concluded that the international community should not fear a relaxation of safety standards in the event of the liberalization of ownership and control of international airlines. 'Ownership' and 'safety' remain two different issues that have to be resolved separately. Indeed, safety concerns will not be resolved by keeping a restrictive regime on foreign investments, but by taking measures that are directly related to aviation safety and security. Thus, in order to ensure a maximum level of safety worldwide and to avoid any conflict between different national safety regulations,[6] the harmonization

---

between the Netherlands and the US on the grounds that KLM would no longer be under Dutch control and therefore in violation of the 1992 US-Netherlands Agreement', see WTO, *Note on Developments in the Air Transport sector Since the Conclusion of the Uruguay Round, Part Four.* WTO Doc. S/C/W/163/Add.3 (2001) 20 [hereinafter WTO doc.2].

[3] The US did not object to Swissair taking control of Sabena, *inter alia*, because the US had Open-Skies agreements with both countries and (therefore) no restrictions on US carriers' traffic rights *vis-à-vis* both countries.

[4] The seventh freedom is the fifth freedom without the requirement that the flight begins or ends in the territory of the contracting party designating the air carrier.

[5] See Part 2, Chap. 3, para. 3, above.

[6] 'The most likely foreign investors in US domestic airlines would be foreign airlines governed by their homeland safety regulations. One can easily envision a non-US owner challenging US safety regulations that differed from those of the foreign owner's homeland, where the foreign owner believes its homeland regulations to be superior. (...) This in turn could create practical difficulties as influential foreign owners could bring political and/or diplomatic pressure to bear on US authorities to either amend or permit variances from US safety regulations', see IATA, Government and Industry Affairs Department, *Report of the*

of national regulations should be provided by the international community before lifting the foreign investment restrictions.

In addition to the minimum safety standards defined by ICAO (the *Chicago Convention* and SARPs), the Organization has created, in January 1999, the Universal Safety Oversight Audit Programme (USOAP).[7] A total of 177 States have been audited under the USOAP so far and, as of 15 November 2001, 168 corrective action plans had been submitted by the States that were audited. The question to raise is therefore how will ICAO be able to implement the necessary measures that are foreseen in the USOAP. To provide States, in need of assistance, with the necessary resources to implement safety-related projects, and to correct deficiencies identified through the USOAP, the ICAO Assembly endorsed the establishment of an International Financial Facility for Aviation Safety (IFFAS). The IFFAS, that will be in place by June 2003, is to be financed by voluntary contributions from States as well as from non-traditional sources, including contributors within or beyond the aviation community.[8] In addition to global plans, safety actions have been taken regionally as well. The EU has been very active in recent years with respect to aviation safety concerns. On 27 September 2000, the European Commission adopted a proposal for a European Parliament and Council Regulation, which would put in place a Community system of air safety and environmental regulation and would create a single European Aviation Safety Agency (EASA). The EASA would take on the mantle of the Joint Aviation Authorities (JAA), which is currently responsible for developing aviation safety criteria. The proposed Agency will develop its know-how in all the fields of aviation safety in order to assist Community legislators in the development of common rules in the field.[9] In addition, the European Commission has proposed

---

*Ownership & Control Think Tank World Aviation Regulatory Monitor*, IATA doc. prepared by H.P. van Fenema (7 September 2000) at 64 (Separate comments of Joanne W. Young) [hereinafter IATA doc.].

[7] See R.I.R. Abeyratne, *Emergent Commercial Trends and Aviation Safety* (Aldershot: Ashgate, 1999) [hereinafter Abeyratne *Emergent Commercial Trends*]; M. Milde, 'Enforcement of Aviation Safety Standards' (1996) 45 Abhandlungen 3 at 4-9. The program has been very well accepted by States, e.g. by the US; C. Shifrin, 'FAA plans safety change' *Airline Bus.* (June 1999) 11; 'Aviation safety will be strengthened further with the Assembly's approval for expanding the Universal Safety Oversight Audit Programme (USOAP) as of 2004. The Programme consists of regular, mandatory, systematic and harmonized safety audits carried out by ICAO in all 187 Contracting States. Since its creation on 1 January 1999, it has proven effective in identifying and correcting safety deficiencies in areas of personnel licensing and airworthiness and operation of aircraft; it will now include air traffic services, aerodromes and the core elements of accident and incident investigation.', ICAO, Immediate Release (PIO 18/2001), 'Safety and Security at the heart of key ICAO achievements for 2001' (27 December 2001), online: ICAO http://www.icao.int/icao/en/nr/pio200118.htm (date accessed: 1 June 2002).

[8] ICAO, Immediate Release PIO 12/2002, 'Council of ICAO Establishes Global Financing Facility for Aviation Safety' (9 December 2002), online: ICAO http://www.icao.int/icao/en/nr/pio200215.htm (date accessed: 24 December 2002).

[9] For additional information on EASA, see EU, *Commission Proposal NO. 500PC0595 for a Regulation of the European Parliament and of the Council on Establishing Common Rules*

concluding agreements with ICAO, the JAA, and EUROCONTROL.[10] These agreements will form the basis for the European Community assistance to ICAO. They will enlarge its audit programme and provide assistance to the Community through all of these organizations. The main purpose is to secure consistent and coherent Community intervention in the form of remedial projects in countries where audits have revealed deficiencies. At the same time, actions are also being taken with regard to the harmonization of aviation security standards, particularly since the tragic events of 11 September. An ICAO resolution calls for a full review of international aviation security conventions and of Annex 17 of the *Chicago Convention*.[11] Moreover, the Assembly directed ICAO to consider the establishment of a Universal Security Oversight Audit Programme, modelled after the highly successful USOAP, in order to assess the implementation of security-related SARPs.[12] At the European level, on 10 October 2001, the Commission proposed the adoption and enforcement of common EU security rules for civil aviation. The common rules will be based on the rules set out in Document 30 of ECAC and will aim for increased control of both international and domestic flights.[13]

Thus, in the event of the liberalization of the ownership and control of international airlines, the fear about the flags of convenience, whether or not this fear is justified, will no longer be a concern once the harmonization of aviation safety and security measures is achieved globally. Wassenbergh indicates that it is the duty of ICAO to achieve 'a global harmonization of national aviation laws regarding [aviation safety and security] together with an adequate system to monitor the implementation of the rules in practice, under penalty of exclusion

---

*in the Field of Civil Aviation and Creating a European Aviation Safety Agency*, (2 July 2001), online europa
http://www.europa;eu;int/eur-lex/en/com/dat/2000/en_500PC0595.html (date accessed: 15 January 2002); C. Thornton, "The Europe's New Transport Commissioner has Set out her Agenda on Air Transport and Appears Determined to See it Through' *Airline Bus.* (March 2000) 32.
[10] EUROCONTROL is a European body in charge of the development of a coherent and coordinated air traffic control system in Europe.
[11] *Convention on International Civil Aviation*, 7 December 1944, 15 U.N.T.S. 295, ICAO Doc. 7300/6 Annex 17 (about safeguarding International Civil Aviation Against Acts of Unlawful Interference) [hereinafter the *Chicago Convention*].
[12] ICAO Immediate Release, *supra* note 8.
[13] For more information about the European actions on aviation security, see EU 'Towards new rules on aviation security following the attacks' Doc. IP/01/1397 (10 October 2001), online: europa
http://www.europa.eu.int/rapid/start/cgi/guesten.ksh?p_action.gettxt=gt&doc=IP/01/1397 (date accessed: 6 December 2001); EU, 'Air Security – Short presentation of the proposal for a regulation establishing common rules for civil aviation security' Doc. of the European Commission Directorate General for Energy and Transport (October 2001), online: europa http://www.europa.eu.int/common/transport/library/press-kit-surete-en.pdf (date accessed: 6 December 2002).

from participation in international air transport.'[14] Indeed, in the long run, a harmonized safety and security regulation would prevail among the entire community of States. Consequently, if countries continuously harmonize the minimum standards that are required among the jointly owned and controlled international airlines, cross-border investments will even tend to raise the level of safety and security oversight worldwide.

## 4. Urgent Internal Political Changes in the EU and in the US

*4.1 A Desirable Resolution of the Commission/Member States Conflict*

To a global extent, various levels of national ownership and control policies lead to unbalanced international air transport relations, as States become unequal in terms of foreign investments, and indirectly, in terms of market access. A prior consensus between communities of States on their transport policy is a prerequisite to liberalization of the airline ownership and control criteria. To this aim, harmonization of national policies has to be proceed, primarily, on a regional scale. This is, currently, a main challenge of the EU.

The adoption of a common policy of ownership and control, not in terms of level of restrictions, as the restrictions have been clearly defined in the EU,[15] but in terms of airline designation, would allow the Community to act as a truly 'Community of interest' and to strengthen its negotiations with third countries, especially with the US. Europe needs to be one sole voice *vis-à-vis* the strong US market to start working on the liberalization of the ownership and control criteria.

The ECJ's ruling on 5 November 2002 on the Open-Skies cases marks an important step in the harmonization process.[16] As explained above,[17] the Court states that the nationality clauses in the bilaterals is a clear violation of the right of establishment, under Article 52 of the Treaty, and should be replaced by the Community clauses. The notion of 'Community carriers' has been recognized by the Third Package in 1992. Since then, all Community carriers have equal access to international routes from any Member State, regardless of the nationality of their

---

[14] H. Wassenbergh, 'Towards Global Economic Regulation of International Air Transportation through Inter-Regional Bilateralism' The Hague (August 2001) at 11 [Unpublished].

[15] See Part 1, Chap. 2, para. 3, above.

[16] *ECJ, Commission of the European Communities v. United Kingdom of Great Britain and North Ireland* (C-466/98), *Commission of the European Communities v. Kingdom of Denmark* (C-467/98), *Commission of the European Communities v. Kingdom of Sweden* (C-468/98), *Commission of the European Communities v. Republic of Finland* (C-469/98), *Commission of the European Communities v. Kingdom of Belgium* (C-471/98), *Commission of the European Communities v. Grand Duchy of Luxembourg* (C-472/98), *Commission of the European Communities v. Federal Republic of Austria* (C-475/98), *Commission of the European Communities v. Federal Republic of Germany* (C-476/98), online europa http://curia.eu.int/ (date accessed: 5 November 2002) [hereinafter ECJ judgment].

[17] See Part 1, Chap. 2, para. 3, above.

owner. As a result, a Belgium airline doing business in the UK is entitled to all rights and privileges of a British airline. However, bilaterals still exclude air carriers of Member States, not parties to the agreements, from the benefit of national treatment in the host Member States, which is a difference of treatment, considered as discriminatory. According to Article 10 of the Treaty, Member States shall take all appropriate measures to ensure fulfillment of the obligations arising out of the Treaty or resulting from actions taken by institutions of the Community. Accordingly, Member States are now obliged to stop contracting new international agreements, as well as to remove the agreements in force, if they infringe Community law. They have one year to denounce there agreements, by a written notice.

The Community clause requirement, reaffirmed by the ECJ, raises two questions. First, whether the designation clause is the main provision of the bilateral agreements, and if so, in its absence, would the bilaterals be deprived of their purpose? In fact, the market access provisions appear to be the central point of the bilaterals, since the purpose of an Open-Skies agreement is to grant fifth freedom to the designated carriers. Without any provisions on traffic rights, it would lose its real substance. However, the designation clause is directly linked to the traffic rights provisions, it remains essential and inherent to the bilaterals. Thus, in the Open-Skies between an EU State and the US, if the EU State cannot favor anymore its national carrier among Community carriers in their external relations, it will have no more incentive to pursue the bilateral regime and will probably have more to gain from negotiating, through the Community, with third countries. It is clear that an EU/US agreement will not be concluded anytime soon, therefore, in the meantime, Member States have to replace the nationality clause in their individual agreement. How much time will it take for the Member States to denounce the nationality clause? The answer remains uncertain, as, at the time of writing, the Member States have not reacted yet to the Court's request. It will certainly take some times as, linked bilaterally, the defendant States will have to take into account the position of the other party to the agreement, i.e. in most cases the US. This leads to the second concern raised by the ECJ's ruling. How will the US react to the ECJ's requirement regarding the replacement of the nationality clause? The ECJ judgments do not commit the American authorities, and therefore, the US probably analyze carefully the consequences of such a decision, mainly in terms of market access. On 8 November 2002, US officials at the DOT provided some elements of response.[18] The DOT refuses the view that 'Americans have established a huge advantage over [their] European trading partners by retaining these nationality clause in [their] bilaterals' and argues that, first, in the Open-Skies agreements between the EU States and the US, EU air carriers are already allowed to operate from any city in any EU State; and second, that the nationality clause offers the US the possibility to reject a carrier on the grounds that it is not owned and controlled by citizens of that country, but does not oblige the US to do so.

---

[18] J.N. Shane, Association Deputy Secretary, DOT, 'Open Skies agreements and the European Court of Justice' (American Bar Association, Forum on Air and Space Law, Hollywood Florida, 8 November 2002).

Accordingly, the US take a position on a case-by-case basis, and do not automatically reject an airline that does not fit the agreement. The DOT official adds that the US do not 'treat the traditional formula as sacrosanct'. The DOT cites, as an example, the *APEC Agreement*, that uses the 'principal place of business' criterion, to support its position.

In response to the DOT statement, it has to be stressed that the transatlantic market is the most important air transport market worldwide, which makes this market specific and very protected. The US have obviously different interests in all the individual agreements with EU Member States, depending on the type of agreements concluded (i.e. the UK/US agreement is very restrictive while the Netherlands/US agreement is an Open-Skies agreement) and on the size of the markets (i.e. the US have much more to lose market share in the large German/US routes than in the Sweden/US routes). Consequently, the US will accept to combine all their interests and introduce the Community clause if they do not lose any market share in such a huge market. The fear to lose market share comes from the fact that the Community clause will certainly enhance competition within the Community, as Community carriers will penetrate more national markets in the EU. In sum, if there were more uniformity between all the Open-Skies agreements signed with individual Member States, the US would certainly be more eager to introduce the Community clause. For the moment, the diversity of the agreements leaves the implementation of the Community clause uncertain.

The various number of States involved in the modification process of the bilateral agreements does not allow any rapid change. An EU/US agreement is likely the long-term solution. Such an agreement would solve the divergence problems regarding the designation clause, and would allow, on a global extent, to implement the liberalization of the ownership and control restrictions. Indeed, the transatlantic market being the most important air transport market worldwide, once an agreement will be negotiated by the two leveraged parties about foreign investment and traffic right policies, the States presently reluctant to such a regulatory shift, will be encouraged to cooperate and to liberalize their policy if they want to remain competitive on the international market. Nevertheless, an EU/US agreement can be foreseen only after the Council grants the Commission a mandate to negotiate on behalf of the Member States. From 1990, the Commission has submitted to the Council several requests to get a mandate to the aim to negotiate an air transport agreement with the US. In June 1996, the Council granted the Commission a partial mandate, to negotiate with the US, in liaison with a special committee appointed by the Council, in relation to competition rules, ownership and control of air carriers, CRSs, code-sharing, dispute resolution, leasing, environmental clauses and transitional measures. The Council explicitly excluded from the mandate negotiations regarding market access, capacity, carrier designation and pricing. Thus, the ownership and control criterion is expressively divided by the Council into two issues: the airline ownership and control and the airline designation. By this confusing distinction, on the one hand, the Council is willing to provide for the possibility for both parties, i.e. the EU and the US, to invest up to 49 per cent in each other carriers in the first phase and more, later on; and on the other hand, the Council does not allow the Commission to discuss the designation clause, as this allowance risks to give the Commission too much negotiating power regarding the attribution of traffic rights.

Today, the Commission still claims the authority to negotiate directly with the US. In its communication published on 21 November 2002,[19] the Commission numbers four negotiating priorities to be taken for the EC, in terms of opening negotiations.[20] The first and highest priority of the Commission is to enter into negotiations with key bilaterals partners. The Commissions considers that '[i]t is of the utmost importance to open Community negotiations that will offer the opportunity of correcting problems in the bilaterals' and that 'the Council should approve a mandate to negotiations as soon as possible on the basis of the preparations already undertaken addressing not only market access actions but also the regulatory environment'.[21] A transport Council held on 5 December 2002 regarding *inter alia* the consequences of the ECJ Open-Skies judgments.[22] The Council took note of the presentation of the Commission's Communication and keeps on working on the question of establishing a negotiating mandate for the Commission. The President concluded that 'the Member States are prepared to co-operate in a constructive spirit with the Commission with the aim of resolving the situation following the Court's ruling on 5 November'.[23] Although the European Commission does not expect Member States to grant a mandate until it is absolutely required, it is now time for EU Member States to understand this necessary shift in order to pursue and finalize the air transport liberalization: granting the Commission a full mandate is one of the prerequisites to revising the ownership and control principle.

It can be noted that the US are willing to have aviation negotiations with the Community and are opened to all issues for discussion. On 8 November 2002, a DOT official stated that 'an essential prerequisite is a like-minded partner on the one side of the negotiating table that represents an airline industry and an aviation market comparable to our own. The EU airline industry, taken as a whole, and the EU aviation market place, taken as a whole, certainly satisfy that test'.[24] However, the DOT sees the EU/US negotiations as a long-term solution, that will begin only after the conflict resolution between the EU and its Member States.

Thus, because the Council has not granted the mandate yet, and because, even after the Commission gets the authority to negotiate, it will certainly take few years to reach an EU/US agreement, the bilateral regime has to be maintained in

---

[19] EU, Commission Communication on the consequences of the Court Judgment of 5 November 2002 for European Air Transport Policy, COM(2002) 649 final, 19 November 2002.

[20] The four priorities are '(a) To enter into negotiations with key bilateral partners; (b) To continue to build-up relations with neighbouring countries; (c) To build-up relations with developing countries; (d) To assert the position of the Community in multilateral for a and work for reform internationally', *ibid.* at 13.

[21] *Ibid.*

[22] EU, *Council meeting N° 2472 on Transport, telecommunications and Energy*, 15121/02 (Press 380) 5-6 December 2002.

[23] *Ibid.* at 34.

[24] Shane, *supra* note 18 at 7.

the meantime.[25] Will the ECJ judgment lead to short-term changes? The Court considered the airline designation issue, the pricing issue, as well as the CRSs issue, as part of the Community competence. Since these matters are part of the bilateral negotiations, it is recommended that 'for a successful bilateral negotiation, the Community and the Member States co-operate',[26] and that the previous negotiating experience in other sectors, such as international trade, be used in air transport sector, as a framework of such cooperation.[27]

According to the above, the ECJ judgment will have two positive effects, hopefully not on a too long-term. First, on the internal market, it is hopeful that the condemnation by the Court of the 'nationality clause' will kick off a restructuring of the European airline industry through mergers and takeovers. Second, on the international market, by assuming their joint responsibility and by cooperating in the negotiations with third countries, particularly the US, the Commission and the Member States will likely get a much stronger and a more legitimate position on the international scene – prerequisite to start the liberalization of international air transport.

### 4.2 The US DOT and the US Congress: Necessity of Fair Play in International Aviation Policy

The major decisions of the US CAB and DOT since 1958 show that these authorities have shaped US policy by exercising a great deal of discretion. In each case, the ownership restrictions were interpreted in a way that furthered some overarching economic and/or political objective.[28] Thus, the DOT's decisions have varied substantially according to American economic needs and, more importantly, to the needs of the US airline industry. The DOT has taken the right to adapt the ownership and control restrictions to each specific case and, to some extent, has taken the place of the legislator. Despite this unclear position of the DOT, the US Congress remains clearly hostile to change the ownership statutes, and consequently, the prevailing legal regime with respect to the ownership of airlines is very protectionist. Since 1993, Congress has examined several bills to amend the *Federal Aviation Act*'s ownership provision that would allow up to 49 per cent

---

[25] 'In the end, the court's decisions may not mark an end to bilateralism, as the EC believes, but, instead, the start of a long and complicated political process ending in new forms of agreements', see Zuckert Scoutt and Rasenberger, 'European Court says "Bye-Bye Bermuda"' *Aviation Advisor*, Special Edition (6 November 2002).

[26] F., Sorensen, W., van Weert, A.C., Lu, 'ECJ Ruling on Open Skies Agreements v. Future International Air Transport Relations' (2003) Air & Space L. (Incomplete references as, at the time of writing, the article is not published yet).

[27] *Ibid.*

[28] See Part 1, Chap. 2, para. 2, above; 'DOT arguably reached the outer limits of allowable agency discretion when it permitted KLM to buy forty-nine percent of the equity and twenty-five percent of the voting stock in Northwest', see T.D. Grant, 'Foreign Takeovers of United States Airlines: Free Trade Process, Problems and Progress' (1994) 31 Harv. J. Legis. 63 at 70.

foreign voting stock,[29] but none of these proposals has had the necessary political support to become law. Changing US statutes is very difficult. Some argue that this is due to the fact that some regulations, such as the ownership law, have been in place since the very beginning of the airline industry development and any change of these laws could harm the US economy. Others argue that US law cannot be changed due to the risk it may create since the US air market is so huge that any regulatory change, especially regarding the ownership issue, would affect the national air transport market, particularly the US carrier market share. This second reason is certainly the main concern.

Nevertheless, the timing is appropriate for the US to introduce a new regulatory system. The abolishment of national restrictions on foreign investments is not an option for the US Congress. It should even be considered a priority since it would enable the launch of liberalization of foreign investment globally. Indeed, if the majority of States advocates this liberalization and direct their actions towards it, the US would better take a leadership role with respect to this trend in order to maintain its strong position on the aviation market in the long run. But for such an evolution to occur, the US authorities have to agree first on the policy they want to implement: the Congress should change the US protectionist statutes and the DOT should stop considering ownership and control cases individually and should respect, more objectively, the letter of the law. Hopefully, when the Congress will accept to liberalize the statutes by increasing the ownership limits, while still protecting the US interests, such legislative shift will enable the DOT to play more fairly on the international scene.

## 5. Harmonization of Competition Laws

Earlier in the study, the argument was advanced that it is necessary to harmonize competition rules in the international air industry.[30] Two recent cases in the aircraft manufacturer industry clearly demonstrate the differences among national systems, especially between the EU and the US competition laws,[31] and illustrate the need to

---

[29] The most recent proposal to change the law came in 1995, *Bill to Amend Title 49, United-States Code, to Authorize the Secretary of Transportation to Reduce Under Certain Circumstances the Percentage of Voting Interests of Air Carriers Which are Required to be Owned and Controlled by Persons Who are Citizens of the United-States*, 104th Congress 1st Session, H.R. 951 (15 February 1995), Bill introduced by representative Clinger.

[30] See Part 2, Chap. 4, para. 2, above.

[31] The Boeing/McDonnell Douglas merger in 1999, K. Luz, 'The Boeing/McDonnell Douglas Merger: Competition Law, Parochialism, and the Need for a Globalized Antitrust System' (1999) 32 Geo. Wash. J. Int'l L.& Econ. 155 at 155; T. O'Toole, 'The Long Arm of the Law' – European Merger Regulation and its Application to the Merger of Boeing & McDonnell Douglas' (1998) 11 Transnat'l L. 203 at 203; P.S. Dempsey, 'Competition in the Air: European Union Regulation of Commercial Aviation' (2001) 66 J. Air L. & Com. 979 at 1117-1122 [hereinafter Dempsey 'Competition in the Air']; the 2001 General Electric Co. proposal to takeover its rival Honeywell, see F. Guerrera, 'GE fires salvo at European Commission' *Financial Post* (5 November 2001) FP11; P. Bocev, 'l'Europe recale le

develop an effective system for reconciling the differences in countries' domestic laws and politics. The airline industry is concerned about the conflict of laws, which may impede the industry growth. In order to prepare the effects of the ownership regime liberalization, in the light of international mergers and takeovers among air carriers, a uniform regime of antitrust laws would be welcomed.[32]

However, unlike the harmonization of safety standards, the harmonization of national competition laws is much more complex. The achievement of multilateral cooperation will only come if the airlines and their governments are prepared to work together for mutual long-term advantage through compromise. However, serving all the national and international private interests, which are so different from each other, within one agreement would be a very ambitious task. Indeed, since the 1990s, there have been a number of attempts at harmonization, which only confirm the difficulty in reaching consensus. The Munich Group in the 1993 GATT Plurilateral Agreement proposed the creation of a code of international antitrust law and an international enforcement agency to address disputes as they arise.[33] Despite the advantage of uniformity and clarity of such a system, it has not been possible to implement it yet, mainly due to the fact that, even if an international code acceptable to all countries could be developed, disputes in interpreting these laws would inevitably arise.[34] Aside from this international initiative, the bilateral system has been used to propose certain initiatives with respect to harmonizing competition law. In that respect, the US has twice attempted to negotiate bilateral agreements encompassing antitrust issues. In 1991, the EU and the US reached an understanding with regard to the implementation of their respective laws.[35] This agreement was intended to promote coordination between the EU and the US to reduce the danger of differences in their respective competition rulings concerning transatlantic mergers. The agreement achieved little success owing to the absence of an agreement for the exchange of confidential information.[36] The second unsuccessful attempt was enacted in the *International Antitrust Enforcement Act* of 1994 (IAEAA),[37] which

---

mariage GE-Honeywell' *Le Figaro* (4 July 2001) 1; A. Chuter, 'Growing pains: Differing approaches taken by US and European regulators over the proposed GE/Honeywell merger highlight a need for common guidelines' *Flight Int'l* (19 June 2001).

[32] A.C. Lu, International Airline Alliances : EC Competition Law/US Antitrust Law and International Air Transport (The Hague: Kluwer Law International, 2003) at 295; G.P. Elliott, 'Antitrust at 35,000 Feet: the Extraterritorial Application of United States and European Community Competition Law in the Air Transport Sector' (1997-1998) 31 Geo. Wash. J. Int'l L. & Econ. 185 at 196-204.

[33] Luz, *supra* note 31 at 171, note 154.

[34] *Ibid.* at 172.

[35] *Agreement Between the Government of the United-States of America and the Commission of the European Communities Regarding the Application of their Competitive Laws*, [1995] O.J. L. 95/47.

[36] For more information about the 1991 Agreement, see Dempsey 'Competition in the Air', *supra* note 31 at 1102-1103.

[37] *International Antitrust Enforcement Act*, 15 U.S.C. §§ 6201-6212 (1994) (hereinafter IAEAA).

authorized the Federal Trade Commission to pursue reciprocal agreements with foreign antitrust enforcement agencies.[38] Furthermore, an additional effort with respect to harmonization was made in 1998 when the EU and the US entered into a supplemental arrangement on the subject of their competition laws: '[I]t was intended to further clarify the principles under which the parties cooperate to eliminate anti-competitive activities in each other's respective territories.'[39]

These few examples of negotiations demonstrate that it is unlikely that an international system of uniform antitrust laws will be achieved soon. At the same time, it represents a necessary complement to the successful implementation of the revision of the ownership regime, since it would be required in order to strengthen the progress of the international airline concentration process. It should therefore remain a priority for the international community. The harmonization approach presents the most viable option for addressing international antitrust concerns, much more than a globalized antitrust law regime, as each State would maintain its own antitrust provisions. The WTO is probably the best entity to be in charge of dispute resolution regarding competition issues, given its great experience in this field and its efficient dispute settlement mechanism.[40]

## 6. Conclusion

This fifth chapter has tried to list the major initiatives that the international community should take within the next years. The implementation of the liberalization of international airline ownership and control may cause problems in the industry if States do not enhance international cooperation. A variety of measures should be taken, and only the principal ones are listed above. Most of these measures would require long and complex negotiating processes. Nevertheless, it is the only way to safeguard the safety and security of the transport worldwide and to 'normalize' the airline industry itself, by allowing it to grow as any other international industry. If the international community intensifies its efforts of liberalization and harmonization, it may reach the final settlement of the ownership issue, which will undoubtedly be subject to a long negotiating process as well. Thus, the sixth chapter of this analysis addresses the intermediate steps to be taken by States before they finalize the liberalization of cross-border investments for the international airline industry.

---

[38] For more information about IAEAA, see Luz, *supra* note 31 at 173.

[39] Dempsey 'Competition in the Air', *supra* note 31 at 1104.

[40] Lu, *supra* note 32 at 289; Luz, *supra* note 31 at 174-177.

Chapter 6

# Regionalism: a Prerequisite to Reach Multilateralism

## 1. Introduction

While the evolution of air transport follows the economic trend towards globalization, the macro-economic dimension has replaced the micro-economic level. Indeed, the national or bilateral negotiating process between States has moved in recent years towards a regional negotiating process en route to a multilateral system. At present, therefore, international air carriers are not evolving in isolation, they are increasingly linked together through bilateral, regional, or multilateral agreements and in this manner they adopt common economic measures. Subject to international negotiations, the ownership and control issue is going to be addressed and resolved in the first instance through a regional approach.

## 2. A Regional Approach to the Airline Liberalization Process

### 2.1 The Role of Regionalism in the Aviation Industry

'The world airline industry has too many players and must face the same painful consolidation that is sweeping other sectors of the economy.'[1] A regional bloc approach is necessary for the consolidation of a liberalized aviation industry, which is 'over-fragmented' at present.[2] Indeed, the airline industry needs a new negotiating process that will foster competition across a broader range of markets. A shift away from bilateralism will likely be spawned by the consolidation that is now underway: with the passage of NAFTA, the development of the ASEAN region as an economic bloc,[3] the completion of the European single aviation market, and the emergence of similar regional affiliations in Central and South America, the potential is there for groups of Nations to negotiate regionally,

---

[1] R. Gibbens, 'World has too many airlines: IATA boss' *National Post* (4 December 2001), online: national Post http://www.nationalpost.com/financialpost/worldbusiness/story.htm (date accessed: 5 December 2001).
[2] Jeanniot Sees Regional Blocs As Cure For Over-Fragmentation, *Aviation Daily* 344:43 (31 May 2001) 2 [hereinafter Jeanniot].
[3] G.C. Hufbauer, C. Findlay, Flying High – Liberalizing Civil Aviation in the Asia Pacific (Washington DC: Institute for International Economics, 1996) 109.

offering their airlines a number of opportunities in accordance with the range of markets. Today, there would be more than 50 different groupings of States that are, or could become, involved in the aviation regulation.

Regionalism presents advantages for developing countries as well as developed countries. It represents an opportunity for developing countries to strengthen their positions in the global air transport market. Indeed, instead of maintaining isolated markets, smaller countries from the same area of the world could gather by common interests – hence the idea of a 'Community of Interests' - and cooperate in order to become a strong community, able to compete with other stronger economic regions. For instance, a Community among Arab airlines is proposed for a single aviation market in the region as they struggle with their weak economic situation.[4] It is likely that '[i]n the long-run, airlines from this region will merge for their own benefit.'[5] And it has been predicted for some years that 'the Latin American airline industry will be consolidated radically over the next five to ten years through large mergers or through the formation of holding companies (...).'[6] Meanwhile, the Andean Pact nations (Bolivia, Colombia, Ecuador, Peru, and Venezuela) agreed to granting each other all five freedoms of the air and to designate multiple national airlines to serve any of the destinations within the region.[7] Regionalism benefits the developed countries as well. The European market is the best model of regional integration in the air transport sector. The 'Three Packages' of liberalization have created a strong 'Community of Interests', able to compete more equally with the US. APEC is another example of the development of regionalism and of its ability to strengthen the regional economy.[8]

Regionalism is valuable for two reasons. First, it allows like-minded States to find initial compromises, which suggests that a multilateral compromise may be easier to arrive at later. Second, it benefits developing countries because, individually, they may not be able to negotiate effectively on market access with dominant nations like the US, but as a regional group they will be able to reap the benefits of a stronger negotiating position enabling their (joint) airlines to compete on more favorable terms in the international market place.[9]

---

[4] 'With the Middle East peace dividend yet to materialize, weak oil prices, a continuing UN economic embargo on Iraq and political instability in many countries, these are challenging times for the Arab world. (...) Arab airlines have not escaped the suffering?', T. Gill, 'Opening Arab Skies' *Airline Bus.* (June 1999) 47.

[5] D. Cameron, 'Out of the Wilderness' *Airline Bus.* (June 1999) 50; Arab cooperation has already started by initiatives in the fields of ground handling and computer reservation system, T. Gill, 'A Firmer Base' *Airline Bus.* (June 2000) 49-50.

[6] C. Shifrin, 'Towards Unsettled Skies' *Airline Bus.* (June 1999) 87.

[7] R.D. Lehner, 'Protectionism, Prestige, and National Security: The Alliance Against Multilateral Trade in International Air Transport' (1995) 45 Duke L.J. 436 at 471.

[8] N. Ionides, 'Spoiling for Choice' *Airline Bus.* (October 2000) 84.

[9] 'Optimistically, the long-term results of a 'plurilateral' relationship could place these [developing countries] in a more favorable negotiating position with the US to finally discard the old bilateral agreements', see G.L.H. Goo, 'Deregulation and Liberalization of Air Transport in the Pacific Rim: Are They Ready for America's 'Open Skies?' (1996) 18 U. Haw. L. Rev. 541 at 562.

## 2.2 Regionalism: a First Step to Liberalize Airline Ownership and Control

Regional agreements could be a bridge to full liberalization, as both de Palacio and Jeanniot assert. Such agreements could be 'progressively extended', with the accelerated elimination of 'artificial barriers to access and entry to markets,' as well as lifting limits on transnational investment.[10] Indeed, inside a 'Community of Interests', States should not fear foreign investment liberalization in their national airlines, if all the measures taken are reciprocal measures agreed upon in the regional agreement. In this case, the traditional concerns regarding the ownership liberalization, such as traffic rights, national security, and economic security, do not seem as relevant, since the regional partners would be equal. Among all liberal regional agreements that have been negotiated so far,[11] two examples, dealing, *inter alia*, with the relaxation of the ownership rule, may be recalled. Among the EU Member States, for example, the traditional national ownership and control requirement was replaced by the concept of 'Community Carrier.'[12] Carriers that meet the legally defined requirements enjoy Community status and can thus benefit from the advantages of Community legislation, e.g., the right of establishment throughout the Community and complete market access, including cabotage. The other agreement, signed on 1 May 2001, which deals with the ownership and control principle of designated airlines is the *APEC Agreement*: more than simply a regional agreement, the *APEC Agreement* is a plurilateral agreement.[13] The Agreement retains the traditional requirement that an airline be 'effectively controlled' by nationals of the country, the government of which designates the

---

[10] Jeanniot, *supra* 2 at 2.

[11] States from different regions of the world have implemented a regional Open-Skies policy, with a liberalization of the ownership and control restrictions as well as of the third, fourth and fifth freedom rights. For instance, the Southern African Development Community (SADC) States are in the negotiating process in order to implement such a policy, see K.J. Max, 'South African liberalisation makes progress' *Flight Int'l* (October 24, 2000) 4; as well, three regional agreements containing 'Community of Interest' provisions have been listed by ICAO (CARICOM, COMESA, ACAC), see WTO, *Note on Developments in the Air Transport sector Since the Conclusion of the Uruguay Round, Part Four*. WTO Doc. S/C/W/163/Add.3 (2001) 25 [hereinafter WTO doc.2].

[12] The three Packages of the European Liberalization, see Part 1, Chap. 1, para. 1, above.

[13] On 15 November 2000, the five partner States reached an agreement, called *Multilateral Agreement on the Liberalization of International Air Transportation* [Hereinafter *APEC Agreement*]; the Agreement contemplates that other countries may sign on and thus provides a potential foundation for a broad multilateral Open-Skies regime. It may be more appropriate to talk about a plurilateral agreement instead of regional agreement, as the *APEC agreement* is much broader than the Asia Pacific region; 'Latin American countries – particularly the Central American countries, where the Group TACA carriers are based – are likely candidates, [M. Gerchick as former DOT deputy assistant secretary] said, pointing to the Bush administration's focus on the Western Hemisphere and well as its commitment to effecting the Free Trade Area of the Americas as a strong reason for extending APEC to other regions', see 'APEC Multilateral Moves U.S. Toward Globalizing Pacts' *Aviation Daily* 344:22 (1 May 2001) 3; N. Ionides, 'Five Sign Up to Asia-Pacific Multilateral Agreement' *Airline Bus.* (June 2001) 34.

airline to serve another country. However, it is significant that the Agreement does away with the traditional requirement that an airline must also be 'substantially owned' by nationals of the designating country. That requirement has been replaced by the condition that the airline be incorporated in that country and has its principal place of business there.[14] Article 3 of the *APEC Agreement* is ambiguous as to whether the ownership and control issue has been liberalized or not. While some Partner States, such as New Zealand, call the international community for a total easing of restrictions in the air service agreements, whether regional or multilateral,[15] other States, especially the US, remain protectionist as regards the market access issue and the ownership and control issue. Furthermore, Article 3 contains an oddity: it states that each government has the right to reject an airline designated by another government if the airline is 'substantially owned' by nationals of the first country. In other words, the US may refuse to authorize service by a Chilean airline if the airline is 'substantially owned' by US nationals. Why would a government want special protection from its own citizens? Why did experienced negotiators allow this very un-liberal provision into their pioneering

---

[14] Article 3 'Designation and Authorization' of the *APEC Agreement*:
  1. Each Party shall have the right to designate as many airlines as it wishes to conduct international air transportation in accordance with this Agreement and to withdraw or alter such designations. Such designations shall be transmitted to the concerned Parties in writing through diplomatic or other appropriate channels and to the Depositary.
  2. On receipt of such a designation, and of applications from the designated airline, in the form and manner prescribed for operating authorizations and technical permissions, each Party shall grant appropriate authorizations and permissions with minimum procedural delay, provided that:
      a.  effective control of that airline is vested in the designating Party, its nationals, or both;
      b.  the airline is incorporated in and has its principal place of business in the territory of the Party designating the airline;
      c.  the airline is qualified to meet the conditions prescribed under the laws, regulations, and rules normally applied to the operation of international air transportation by the Party considering the application or applications; and
      d.  the Party designating the airline is in compliance with the provisions set forth in Article 6 (Safety) and Article 7 (Aviation Security).
  3. Notwithstanding paragraph 2, a Party need not grant authorizations and permissions to an airline designated by another Party receiving the designation determines that substantial ownership is vested in nationals.
  4. Parties granting operating authorizations in accordance with paragraph 2 of this Article shall notify such action to the Depositary.
  5. Nothing in this Agreement shall be deemed to affect a Party's laws regulations concerning the ownership and control of airlines that it designates. Acceptance of such designations by the other Parties shall be subject to paragraphs 2 and 3 of this Article.

[15] Ionides, *supra* note 13 at 34.

new Agreement?[16] In addition, the US DOT has established a reporting system that enables the US to monitor the degree of US ownership of foreign carriers and issued an order that imposed upon the airlines (including Lan Chile, Air New Zealand, Singapore Airlines and Royal Brunei Airlines) of four of the Partner States[17] a 30 days' advance notice to the DOT of any transaction in which a US national either increases its ownership of a foreign carrier's stock by 20 per cent or results in a US national owning 40 per cent or more of a foreign carrier's stock.[18] Accordingly, the *APEC Agreement* is more symbolic than practical for many signatories,[19] as the liberal measures on ownership and control in the Agreement are quite narrow.[20] However, the deal may still prove to be significant. It marks the first major initiative of Washington to move beyond the existing Open-Skies policy, leaving the agreements unchanged, but bundling these liberal bilaterals into multilaterals. Thus, it can serve as a starting point for multilateral negotiations in order to proceed to the complete abandonment of national ownership and control restrictions.

To complete this discussion of regionalism, it bears noting that, as with regionalism in trade, the regional fragmentation of the overall aviation market could result in more competition within specific blocs. However, barriers remain among the blocs, as competitors from outside the agreement cannot compete equally with carriers from inside the bloc. In fact, regionalism is just a starting point towards multilateralism. The different regional groups that are created will need to gather in the long run in order to harmonize, *inter alia*, the cross-border investments among the international airlines around the world. On this point, de Palacio states that 'it has not been possible yet to create links between these

---

[16] To these questions, the answer can be summarized in one word: labor. Indeed, it has been argued many times, as we have seen, that US labor interests were reportedly concerned that cross-border investment would mean that US airlines would export capital and jobs to low-cost, foreign airline subsidiaries. Therefore, US unions reportedly secured a commitment from the US DOT that this fear would be fully addressed. US negotiators then proposed the provision, which was nominally included in the Agreement as a means to address concerns about 'flags of convenience'.

[17] No comparable requirement is imposed on the airlines of any other countries, including countries with the most restrictive bilateral agreements with the US.

[18] In January 2001, according to its own terms, the DOT issued a 'Show Cause Order', 'amending the US licenses of carriers of the participating countries, requiring them to notify DOT of any increase in beneficial shareholding by a US shareholder by 20 per cent or more, within 30 days after such a change', Department of Transportation, *Order in the Matter of Amending the US Licenses of Carriers of the Participating Countries*, DOT Order 01-1-13 (16 January 2001); 'Multilateral Pact Carriers Must Report Ownership Changes' *Aviation Daily* 342:43 (1 December 2000) 4.

[19] This is because most of them already have bilateral Open-Skies agreements with a number of other signatories and some with all of them. Those that do not, such as Singapore and Chile, are not going to see direct flights as a result of this new agreement.

[20] 'Asian carriers have given a cautious welcome to the deal, saying true liberalization will not take place until airline ownership rules are revamped globally', see D. Knibb, 'Bilateral Accord Sparks Ownership Debate...as APEC Moves Towards Multilateral Open Skies' *Airline Bus.* (January 2001) 24.

regions and thus allow airlines to benefit from a much larger market, but the Commission is actively pursuing this goal and tries to develop relations with these other regions.'[21] Since 1995, attempts have been made to get the two main aviation regions of the world, the US and the EU,[22] together in a so-called Transatlantic Common Aviation Area (TCAA). Proposed by the Association of the European Airlines (AEA), the TCAA contains virtually all of the features of the US model Open-Skies agreement, but goes considerably beyond that model.[23] Briefly, in addition to allowing full pricing freedoms and providing alliances with operating flexibility, the TCAA identifies four core areas for liberalization: first, the freedom to provide services between any points in the Area, including two points in a single country (cabotage); second, unrestricted airline ownership and the right of establishment;[24] third, the harmonization of standards for the evaluation of airline competitive behavior;[25] and fourth, the elimination of restrictions on the use of leased aircraft and on the reservation of the carriage of government-financed traffic to national carriers.[26] The TCAA proposal seeks to address the problem of the lack of government policy coordination by creating a mechanism for regulatory convergence. Indeed, it is a priority to establish a dialogue between the EU and the US, about their respective visions of the market-based regulatory environment in which they want airlines to operate.[27] Even though some people remain opposed to such an opening of the EU/US air transport market,[28] the TCAA would benefit the European airline industry as much as the US airline industry, the consumers and the shareholders.[29] However, there have been only informal talks between Brussels

---

[21] L. de Palacio, 'Globalization – The way forward' (IATA World Transport Summit, Madrid, May 27-29, 2001) [hereinafter de Palacio 2001].

[22] 'We are all well aware of the fact that the existing bilateral air transport relations form a complex political web. It will not be easy to replace. Therefore the Commission is of the opinion that the two most important air transport markets in the world, the US and EU should take the lead', see L. de Palacio, 'Beyond Open Skies' (the European Commission Beyond Open Skies Conference, Chicago, 6 December 1999).

[23] Wassenbergh proposes a Draft text of a supplemental bilateral air transport services agreement between Member States of the EU and the US, see H. Wassenbergh, 'Towards Global Economic Regulation of International Air Transportation through Inter-Regional Bilateralism' The Hague (August 2001) at 19 [Unpublished].

[24] Ownership and control of carriers could be vested in nationals of any of the participating TCAA Member States.

[25] R.I.R. Abeyratne, 'Emergent Trends in Aviation Competition Laws in Europe and in North America' (2000) 23 World Competition. R. 141 at 154 [hereinafter Abeyratne 'Aviation Competition Laws'].

[26] For an overview of the AEA's proposal: AEA, *Towards a Transatlantic Common Aviation Area*, AEA Policy Statement (September 1999).

[27] The European Commission's vision of a regulatory structure that will allow airlines to benefit from liberalization, see de Palacio 2001, *supra* note 21.

[28] Especially the transport worker unions, see B. Lancesseur, 'Un Cadre Réglementaire rigide – La mise à plat s'impose' *Aéroports Magazine* (Mai 2001) 18.

[29] U. Schulte-Strathaus, 'Common Aviation Areas: the Next Step Toward International Air Liberalization' (2001) 16-SUM Air & Space L. 4 at 5; European carriers are determined to push the TCAA project, see M. Pilling, 'Only a Call Away' *Airline Bus.* (March 2001) 39;

and Washington so far, and although the political negotiations will continue, no one seems optimistic about a transatlantic breakthrough being achieved soon. First, too many disagreements still remain between the parties. 'A TCAA may be agreed upon only if it is limited to the exchange of third and fourth freedoms and, for EU air carriers, seventh freedom air traffic rights (not 'external' fifth freedom),'[30] and only if the ownership and control liberalization is limited to 'the APEC formula for ownership of designated air carriers,'[31] in accordance with the prevailing US view. Second, external factors to the TCAA impede further negotiations. The US/UK relations regarding air transport liberalization are still unclear. The UK, often considered as the 'champion' of liberalization and free competition, continues to protect access to London Heathrow airport, when almost all other EU Member States have an Open-Skies agreement in place. At the present time, negotiations continue between the two States but it is still doubtful whether an Open-Skies will be concluded anytime soon, particularly in view of the recent ECJ judgment. The other external obstacle to TCAA negotiations is Ireland that still protects the so-called Shannon stop for reasons of domestic economic policy. Furthermore, the default of the European Commission's mandate remains the biggest hurdle that stands in the way of the TCAA negotiations. In its final judgment as regards the Open-Skies cases,[32] the ECJ confirmed that the claimed exclusive competence of the Community to conclude an agreement with the US is not founded.[33] Although it is time for EU Member States to understand this necessity to grant the Commission a mandate in order to pursue and finalize the air transport liberalization, the Council has not decided yet to grant such a mandate, and it will probably take some time before a change occurs in the Member States' positions, since it is not expected that they grant a mandate to the Commission before it is absolutely necessary.

Considering all of the above, the process of cross-border investment liberalization seems arduous, even at the regional level. As long as the negotiations remain between States, as part of the same 'Community of Interests', the

---

Air France calls for further negotiations on TCAA, see Lancesseur, *ibid.* at 21; C. Baker, 'French Push for TCAA' *Airline Bus.* (December 2000) 18.

[30] Wassenbergh, *supra* note 23 at 9.

[31] Otherwise 'it is likely that [the US] Congress would have to amend the *Federal Aviation Act* before the US could sign the TCAA', see *ibid.*, see also Schulte-Strathaus, *supra* note 29 at 6.

[32] *ECJ, Commission of the European Communities v. United Kingdom of Great Britain and North Ireland* (C-466/98), *Commission of the European Communities v. Kingdom of Denmark* (C-467/98), *Commission of the European Communities v. Kingdom of Sweden* (C-468/98), *Commission of the European Communities v. Republic of Finland* (C-469/98), *Commission of the European Communities v. Kingdom of Belgium* (C-471/98), *Commission of the European Communities v. Grand Duchy of Luxembourg* (C-472/98), *Commission of the European Communities v. Federal Republic of Austria* (C-475/98), *Commission of the European Communities v. Federal Republic of Germany* (C-476/98), online europa http://curia.eu.int/ (date accessed: 5 November 2002) [hereinafter ECJ judgment].

[33] *Ibid.*, see, *inter alia*, Case C-476/98 Commission of the European Communities v. Federal Republic of Germany, para. 111.

liberalization process works quite well (e.g. the EU), but as soon as regional agreements broaden and become plurilateral, such as the *APEC Agreement*, liberalization is inevitably more limited since various national interests and policies, sometimes conflicting, have to be taken into account. Thus, the harmonization of regional policies regarding, *inter alia*, the ownership and control of airlines, is much more complex. However, through a progressive expansion of all the regional agreements, in the long-run, a complete liberalization of foreign investments worldwide should be achievable.[34]

## 3. A Multilateral Approach to the Airline Liberalization Process

### *3.1 The Role of Multilateralism for the Airline Industry*

A multilateral framework would enable the international community of States to harmonize their air transport regulations. While some argue that multilateral liberalization will be reached only through a progressive expansion of the various regional agreements (e.g. the TCAA), others believe that a multilateral agreement can be achieved in a few steps under the auspices of an international organization such as ICAO. Even though the importance of the role of international organizations with respect to air transport liberalization should not be disregarded,[35] it is likely that a multilateral agreement that harmonizes air transport rules, such as the airline ownership and control issue, will be reached by harmonizing the regional agreements incrementally. This process will take time. Thus, with the air transport world divided into regions, a multilateral framework 'may be achieved by liberal 'bilateralism' between regions. It will not be in the world's interest, if the regions (...) become ever so many protectionist 'fortresses' and eventually will start a fight bilaterally between them in order to try and conquer the world's sky, just as the States initially did and still may try to do.'[36]

---

[34] Kass expresses this idea very clearly by stating:
> A plurilateral arrangement will not open all markets immediately, but it will gradually achieve this goal by increasing the number of countries that are committed to a specific set of principles. For example, the US position is that it is unwilling to discuss changes in foreign ownership and cabotage, except on a case-by-case basis...By signing Open-Skies agreements with individual nations, a base of common agreement would be solidified. It would then be much easier to convene a multilateral convention among these countries to consider subjects such as foreign ownership and cabotage.

See H.E. Kass, 'Cabotage and Control: Bringing U.S. Aviation Policy into the Jet Age' (1994) 26 Case W. Res. J. Int'l L. 143 at 180.

[35] The role of international organizations in the process of air transport liberalization will be addressed in Part 3, Chap. 7, below.

[36] Wassenbergh, *supra* note 23 at 11.

Nations and airlines would benefit from a multilateral liberalization agreement. Among other advantages,[37] multilateralism is a better vehicle than bilateralism for achieving widespread liberalization and obtaining the ensuring benefits. It may end the waste of time and the expense of negotiating and renegotiating thousands of bilateral air agreements; it may further promote increased standardization of the numerous regulatory arrangements that exist in the bilaterals; increased international traffic would add to the value of export services and reduce the cost of foreign airline services; and it may 'open developing countries to increased air service by foreign providers, thereby resulting in greater economic development.'[38] Therefore, it would be appropriate to undertake the first step towards the establishment of a multilateral agreement. The question is how a multilateral agreement on air transport could be concluded taking into account the various interest groups concerned, particularly consumers, airlines, governments, and labor.

## 3.2 A New Multilateral Agreement on Air Transport

The international community is facing a very ambitious task regarding the establishment of a multilateral framework on air transport. So far, no international agreements have been reached on the commercial aspects of civil aviation – it was one of the main flaws of the *Chicago Convention* – owing to quasi-global State protectionism and the persuasive use of market access as a bargaining chip. The principal objective of such an agreement would be air transport liberalization and it should therefore be designed to meet the needs and serve the interests of the various stakeholders (e.g. air transport consumer and workers, airlines and airports, tourism, trade and related interests). Moreover, a question that is fundamental to the design of the new agreement is whether it would replace existing bilateral agreements or would complement them. Considering the restrictive provisions of the current bilaterals, it is hard to imagine how they can be combined with a multilateral agreement on air transport liberalization. Therefore, in the long run, an exclusive application of the multilateral agreement is advocated.

The structure of this proposed new air transport multilateral agreement should include several essential elements. The provisions must be clear and flexible to permit air carriers to maximize their opportunities. A Preamble should, after defining the principal terms of the agreement, affirm the main goals the Parties want to pursue by implementing the agreement, including: 'promote the freedom of movement of persons, goods, capital, services, ideas and information'; 'maintain a global network of safety...and promote security'; 'optimally advance the solidarity'; and 'meet the needs of international inter-course, trade and tourism.'[39] In addition, the Preamble should guarantee 'an increasing participation

---

[37] 'The merits of a multilateral system' are clearly addressed by Lehner, *supra* note 7 at 458-462.

[38] For more details about this last benefit, see *ibid.* at 466.

[39] Wassenbergh has drafted a proposed Preamble of a multilateral economic air treaty, see Wassenbergh, *supra* note 23 at 11.

of developing countries by taking particular account of the serious difficulty of the least-developed countries' and recognize 'the international public interest as of primary importance,'[40] as well as 'consumer protection'. Finally, the *Chicago Convention* should be included in the Preamble in order to preserve the basis air transport principles, such as the sovereignty of airspace and the equality of opportunity in international air services.

Second, the structure of the multilateral agreement should include two parts. The first part should combine the main economic provisions that were previously approved on a regional basis. Among other provisions, it should deal with 'market access', 'airline designation and authorization', 'safety and security' and 'dispute settlement mechanism'. Regarding the 'market access' issue, the agreement should contain fifth and seventh freedom traffic rights for all contracting parties.[41] Since further liberalization of market access is conceivable in the long run, it would include the eighth[42] and the ninth[43] Freedoms. In fact, as many countries will be hesitant to go as that far, a compromise can foresee that, while the first seven Freedoms are granted directly, cabotage rights may be included subject to reservations. For instance, cabotage rights can be granted only to some of the airlines so designated in the present agreement or only to the air carriers that have entered into an alliance with an authorized designated air carrier. However, this compromise is not the ideal solution as such difference of treatment between air carriers would immediately affect the right of establishment and the ownership and control issue. With respect to the 'designation and authorization' provision, it has to be determined whether it is more appropriate to adopt a provision similar to Article 3 of the *APEC Agreement*,[44] which removes the ownership restrictions, but retains the control requirement of the air carrier by nationals of the designating Party, or to remove both the ownership and the control provisions. Considering our previous argumentation in favor of a total elimination of foreign investment

---

[40] *Ibid.* at 12.

[41] According to the model of Article 2.1 of the *APEC Agreement* (see the *APEC Agreement, supra* note 13) the provision should state:

  Each Party grants to the other Parties the following rights for the conduct of international air transportation by the airlines of the other Parties,
  a.  the right to fly across its territory without landing;
  b.  the right to make stops in its territory for non-traffic purposes;
  c.  the right, in accordance with the terms of their designations, to perform scheduled and charter international air transportation between points on the following route:
  i.      From points behind the territory of the Party designating the airline via that Party and intermediate points to any point or points in the territory of any other Party and beyond;
  ii.     Between the territory of the Party granting the right and any point or points.

[42] The eighth Freedom is the right to carry passengers within a country by an airline of another country on a route with origin/destination in its home country.

[43] The ninth Freedom is the right to carry passengers within an airline's home country.

[44] The *APEC Agreement*, Article 3, *supra* note 14.

restrictions, the second option is advocated. Though it is necessary to keep a strong link between the designating State and the airline for safety reasons. The elements of a strong link include the airline being established and having its principal place of business in the relevant State, and also holding an Air Operator's Certificate from that State.[45] Moreover, the multilateral agreement should refer to Article 4.1 of the *APEC Agreement*, as it gives the Parties more freedom to designate airlines. According to these few remarks, the 'Authorization and Designation' provision of the multilateral agreement can be drafted on the model of Article 4 of the *APEC agreement*, apart from the part 2.a of the Article.[46] As for 'safety and security', the multilateral agreement should guarantee that the level of international air transport safety and security would not be affected by the liberalization. Once again, the *APEC Agreement* can be taken as an example, whereby the contracting Parties agree to 'maintain and administer safety standards' (Article 6) and 'reaffirm that their obligation to each other to protect the security of civil aviation against acts of unlawful interference forms an integral part of this agreement' (Article 7). Nevertheless, these 'promises' do not constitute a sufficient safeguard. To ensure flight safety, the further control of qualified actors, producers, and air carriers is required. Therefore, ICAO has to achieve a global harmonization of national aviation laws regarding certification and licensing, authorization of air carriers, airport operations and air navigation, aviation communication and surveillance while creating an adequate system to monitor the implementation of the rules in practice, under penalty of exclusion from participation in international air transport. Incorporating this system in a multilateral agreement is the only way to ensure its efficacy. Finally, the multilateral agreement should provide a dispute settlement mechanism that can solve any disagreement not resolved by a first round of consultations. A flexible *ad hoc* mechanism has been foreseen by the *APEC Agreement*, in its Article 14.

The second part of the multilateral agreement should include other important principles, either used generally in trade agreements or newly developed, in order to improve the equality among the Parties and, therefore, facilitate the conclusion of such an agreement. Three issues will be presented: the fair competition principle and the convergence of competition policy, the transparency principle, and the reciprocity principle. First, the issue of competition must be included in the agreement, in two ways. Competition among air carriers must be executed fairly, and cannot be left to the free forces of the market place. 'It will therefore be necessary to find and agree upon the exact limits of justifiable government intervention... Also, it may be necessary to try and determine the amount of competition in any given market, which achieve optimal results for the

---

[45] This criterion of principal place of business, originally developed by ECAC, has become more important over time, in the bilateral and regional agreements. For more information about this criterion, see WTO doc. 2, *supra* note 11 at 21-25; P.P.C. Haanappel, 'Airline Ownership and Control and Some Related Matters' (2001) 26-2 Air & Space L. 90 at 101-102 [hereinafter Haanappel 'Airline Ownership and Control']; H.A. Wassenbergh, 'The Sixth Freedom Revisited' (1996) 21 Air & Space L. 285 at 291-292.

[46] The *APEC Agreement*, Article 4, *supra* note 13.

consumer.'[47] For this purpose, a specific provision should be drafted.[48] Moreover, special attention must be given to the issue of competition policy divergence. It has been mentioned earlier that harmonizing the different competition regimes should be a priority in order to strengthen the process of international airline concentration and to equalize the State Parties. Accordingly, the inclusion of an additional provision called 'Promotion of Convergence of Competition Policy' is highly recommended. Another important principle to insert in the multilateral agreement is the principle of transparency. A transparency requirement means that State Parties must promptly provide a publication of all relevant rules and regulations, administrative guidelines, and all other decisions, rulings or measures of general application which pertain to or affect the operation of the agreement to each other. It creates clarity and predictability and it is necessary for maintaining good relations among States. The third issue to address is the reciprocity principle. In order to establish equal relations among the partner States, the multilateral agreement should indicate whether there are reciprocal benefits and costs for each of them. Some authors have declared that 'real reciprocity is unlikely and essentially impossible' among States,[49] owing to the huge differences between the air markets. Indeed, there may be some unbalanced advantages; however, in the long run, reciprocity seems to be more and more possible, as regionalism will progressively reduce the differences between nations. Furthermore, the reciprocity principle is related to the MFN basic GATS principle, which requires that a GATS member accords the service suppliers of other members with treatment that is no less favorable than it accords service suppliers of any other country. Including the MFN principle in the multilateral agreement would mean, for instance, that if the US lift its restrictions on ownership and control of airlines *vis-à-vis* the EU, it would have to do so for all the partner States of the agreement. Such balanced relations would surely improve the international air transport globally.[50]

The structure of a new air transport multilateral agreement should finally include some appendixes that deal with specific issues, such as State assistance and slot allocation. Indeed, in order to render possible the implementation of the agreement, States should anticipate particular situations, draft additional regulations, and foresee specific exceptions.

---

[47] Wassenbergh, *supra* note 23 at 28.

[48] A 'fair competition' provision can be drafted on the model of Article 11 of the *APEC Agreement, supra* note 13.

[49] Edwards states that, by the reciprocity principle, the US will always give more than what they will get, as some countries still refuse to concede liberal rights to US carriers (e.g. the UK), see A. Edwards, 'Foreign Investment in the U.S. Airline Industry: Friend or Foe?' (1995) 9 Emory Int'l L.R. 595 at 629-632.

[50] Instead of applying the MFN principle as such, some authors advocate a 'conditional MFN treatment' scenario where only those Parties willing and able to accede to terms of the Agreement would be required to comply, see R.I.R. Abeyratne, 'Would Competition in Commercial Aviation ever fit into the WTO?' (1996) 61 J. Air L.& Com. 793 at 840 [hereinafter Abeyratne 'Competition in Commercial Aviation']. This option does not fit the multilateral agreement goal, which is to liberalize the air transport constraints and to balance the benefits among States, as it will even increase the differences among them.

## 4. Conclusion

This sixth chapter can be concluded with the observation that liberalization of air transport is now underway, with various regional agreements already concluded, and increasing pressure from the various stakeholders to liberalize further. However, the resolution of certain remaining issues will likely be a difficult task, particularly the airline ownership and control restriction. The recent plurilateral *APEC Agreement* is a good illustration of this arduous task and shows how difficult it is to break through the protectionism of some States, e.g. the US. The TCAA negotiations may face the same reluctance to abolish foreign investment restrictions. Despite this concern, it is likely that the liberalization process will continue to develop towards a multilateral agreement. The US are favorably disposed to the idea of a case-by-case liberalization; therefore, by gathering the liberalizing agreements in stages, and by creating a fair, equal, and reciprocal multilateral agreement, there is a good chance that, over time, the main constraints embodied in today's international air transport policy, which affect the health of the industry, such as ownership and control conditions and limited traffic rights, may finally disappear.

# Chapter 7

# The Role of International Organizations

## 1. Introduction

Since the beginning of international civil aviation, States always had to take the initiative in regulating the commercial aspects of international air transport. Indeed, as the *Chicago Convention* was mute with respect to the exchange of market access/traffic rights between commercial air carriers, in 1946, States started to negotiate bilaterally.[1] More recently, the two ICAO Conferences in 1992 and 1994 have shown how difficult it is to reach an agreement on the replacement of the bilateral system by a global liberal multilateral system. Conscious of the need to start the liberalization process, some States therefore concluded regional agreements. Today, it is time to go beyond the regional stage and progressively turn the regional agreements into plurilateral conventions and, later, into a multilateral treaty.

International organizations have a key role to play in this ambitious task. ICAO, WTO, and OECD are the main organizations that intervene in this process by helping governments to tackle the economic, social, and political challenges of the globalized air transport economy. These three organizations are not in competition, as their roles are different. They cooperate in order to improve and accelerate the liberalization process by providing new ideas and debates with respect to difficult issues, such as the airline ownership and control issue.

Thus, this chapter will analyze the proposals of the OECD, WTO, and ICAO regarding the air transport liberalization process and regarding the ownership and control issue in particular, as well as the role they can play in this process.

## 2. The Organization for Economic Co-operation and Development

### 2.1 The Role of OECD in Air Transport

The OECD groups 30 member States that roughly share a commitment to democratic government and market economy. The OECD plays a prominent role in fostering good governance in public service and corporate activity. It helps

---

[1] R.I.R. Abeyratne, 'Would Competition in Commercial Aviation ever fit into the WTO?' (1996) 61 J. Air L.& Com. 793 at 794 [hereinafter Abeyratne 'Competition in Commercial Aviation'].

governments ensure responsiveness with respect to key economic areas by monitoring the sector. Deciphering emerging issues and identifying effective policies, it helps policy-makers adopt strategic orientations. It is well known for its country surveys, reviews, publications and statistics.

The OECD has an important role to play in the air transport liberalization process and in the establishment of a multilateral agreement. So far, the Organization has produced several internationally-agreed instruments, decisions and recommendations, and has provided input to the policy debate on current and emerging issues with respect to the prospect of a multilateral agreement. It has provided forums for all participants in the transport chain to consider the suggested reforms. To coordinate the various interests and to avoid any overlap in activities, the OECD Secretariat has maintained close contact with the AEA, ECAC, IATA, ICAO, and WTO. The most recent and interesting OECD proposal regarding air transport liberalization is its regulatory reform in international air cargo transportation.

### 2.2 'OECD Principles for the Liberalization of Air Cargo'

In 1999, the OECD took an initiative with a potential impact on the air transport liberalization process. In June of that year, the OECD transport division organized a workshop on 'Regulatory Reform in International Air Cargo Transportation'.[2] This initiative was followed, in June 2000, by the preparation of a Working Paper entitled 'OECD Principles for the Liberalization of Air Cargo',[3] and, on 4-5 October 2000, by a further workshop in Paris that identified issues and possible approaches. Following the 2000 Workshop, the OECD coordinated the activities of Informal Working Groups led by representatives of individual Member States – with the participation of governments, international organizations, aviation industry and air cargo – that had been formed to address the issues. The '2000 Liberalization Principles' document has been proposed by the OECD Secretariat to assist the interested parties with the liberalization of air cargo services. It suggests practical ways to promote liberalization in the air cargo transport sector, identifying conceptual issues that need to be addressed and the principles that should guide liberalization initiatives. Then, the document focuses on two alternative broad implementation approaches that may be taken: first, by amending existing bilateral agreements, and second, by introducing a new multilateral agreement. In other words, the first proposal is a Protocol for existing air service agreements that liberalizes certain specific air cargo issues and that would allow early implementation of targeted improvements to air cargo arrangements. The second proposal is a multilateral agreement that could provide an effective and

---

[2] OECD, Directorate for Science, Technology, and Industry – Division of Transport, *Regulatory Reform in International Air Cargo Transportation*, Doc. No. DSTI/DOT(99)1 (June 1999).

[3] OECD, Directorate for Science, Technology, and Industry – Division of Transport, *OECD Principles for the Liberalization of Air Cargo*, Doc. No. DSTI/DOT(2000)1 (June 2000) [hereinafter *OECD Liberalization Principles*].

efficient alternative approach for a body of interested Member States wishing to liberalize market access as well as auxiliary cargo-related services without reliance on re-negotiating a complex web of bilateral air service agreements.[4]

Concretely, the OECD Principles deal, *inter alia*, with 'grant of traffic rights', 'designation and authorization', 'prices', 'consumer protection', 'leasing', 'fair competition', 'promotion of convergence of competition policy', 'safety and security', 'ground handling', and 'dispute settlement'. With respect to the airline ownership and control issue, the OECD proposes the removal of both the ownership and the control requirement, thus going further than the APEC designation clause. Thus, Article 3 of the proposed multilateral agreement, called 'designation and authorization', suggests a multiple designation provision,[5] with a 'principal place of business' criterion. Article 3 has received various comments from States. For instance, France has approved it, stating that 'there is no reason to question the aptness of a system that would subject combined and all cargo carriers to different ownership and control regimes depending on whether or not they subscribed to the multilateral agreement', while Greece has refused the multiple designation provision.[6]

The OECD initiative with respect to the liberalization of air cargo only is based on the view that international air cargo demands are continuing to increase more rapidly than international air passenger demands, and the air cargo constraints 'restrain [the carrier's] *corporate and business structures*, notably their *ownership and control* structures, the *possibility* to contract freely with domestic/local carriers abroad, and to diversify into complementary services such as *freight-forwarding*. Taken together, these constraints prevent air carriers from developing the *seamless* transport services needed by domestic and international

---

[4] For more information about the '2000 Liberalization Principles', see *inter alia* the following documents, OECD, Directorate for Science, Technology, and Industry – Division of Transport, *Draft Annotated Agenda – OECD Workshop on Principles for the Liberalization of Air Cargo Transportation*, Doc. No. DSTI/DOT/A(2000)1 (August 2000); OECD, Directorate for Science, Technology, and Industry – Division of Transport, *OECD Workshop on Principles for the Liberalization of Air Cargo Transportation – Paris, 4-5 October 2000 – Summary Record*, Doc. No. DSTI/DOT/M(2000)1 (November 2000); OECD, Directorate for Science, Technology, and Industry – Division of Transport, *OECD Principles for the Liberalization of Air Cargo Transportation – Comments on DSTI/DOT(2000)1*, Doc. No. DSTI/DOT/RD(2000)1 (September 2000).

[5] A 'multiple designation provision' means that the contracting parties have the right to designate several carriers. The OECD proposes Article 3.1 be drafted as follows: 'Each Contracting Party shall have the right to designate, on a non-discriminatory basis, any number of its air carriers for the conduct of international air cargo transport services in accordance with this Agreement and to withdraw or alter such designations. Such designations shall be transmitted to other contracting parties in writing through diplomatic channels', see Directorate for Science, Technology, and Industry – Division of Transport, *OECD Principles for the Liberalization of Air Cargo Transportation – Comments on Articles contained in DSTI/DOT(2000)1*, Doc. No. DSTI/DOT/RD(2000)2 (September 2000) at 9.

[6] For the State comments of all the provisions, see OECD, Doc. No. DSTI/DOT/RD(2000)2, *ibid*.

customers.'[7] Moreover, it seems that air cargo issues are by nature less subject to national sensitivities, and much fewer political issues are raised than with respect to those related to passenger transport. Therefore, it was easier to start the necessary air transport liberalization process with respect to air cargo for a greater acceptance by the interested parties.

The well drafted '2000 Liberalization Principles' proposal can certainly serve as a basis for the international community of States to create the framework for air transport liberalization. However, although the proposal has received widespread support from governments and from the largest international carriers,[8] an important concern remains given that setting air cargo free is not simple, as about '60 per cent of it still travels in the belly of passenger aircraft: no possibility of that being freed separately from passenger rights.'[9] Therefore, the only way to apply the OECD proposal is to separate completely the air cargo operations from the passenger operations. The document has been submitted to national delegations, organizations and relevant industry parties for comments by the end of 2001. As a result, the industry representatives were strongly of the view that current air transport regulatory arrangements were inappropriate for the needs of air cargo transportation and supported the liberalized approaches outlined by the Secretariat.

Following the 2000 Workshop, a revised paper on liberalization principles and approaches was prepared by the OECD Secretariat taking into account the deliberations of Informal Working Groups that focused on the key issues. Contrary to the previous version of the document, the provisions suggested in Parts III (Bilateral Protocol) and IV (Multilateral Agreement) were restricted to air carriers.[10]

The final OECD Workshop was held on 21-22 January 2002. This Workshop considered that relaxation of existing restrictions on international traffic rights would allow better market access and improve the industry's ability to meet user requirements. Relaxation of government controls over 'ownership and control' of air carriers would allow the development of industry structures and services better suited to global and regional air cargo tasks. There was also support,

---

[7] *OECD Liberalization Principles, supra* note 3 at 3.

[8] P. Conway, 'Could Cargo Lead Liberalisation' *Airline Bus.* (December 2000) 29.

[9] *Ibid.*

[10] 'The reasons for this were basically threefold: *first* including all sorts of air cargo service providers in the scope of an air cargo regulatory regime could add government oversight to a large number of hitherto unregulated services and entities which might not be desirable or in accordance with the scope of authority under national law; *second* for some ancillary services, jurisdictional and/or policy co-ordination among multiple regulatory bodies, where they exist, might prove difficult for various governments and *third* including all air cargo service providers could be a factor substantially delaying the adoption of any such document as this could necessitate extended and, at various occasions, complicated co-ordination among agencies/ministries concerned', see OECD, Directorate for Science, Technology, and Industry – Division of Transport, *Liberalization of Air Cargo Transport*, Doc. No. DSTI/DOT(2002)1/REV1 (May 2002) at 5.

although not consensus, for broadening the scope of the regulatory reform to encompass liberalization of international air cargo services generally, whether the cargo was carried on all-cargo services or on combination services carrying passengers and cargo. There was general agreement that liberalization measures should only be undertaken if essential public interest regulatory controls over aviation safety and aviation security can be assured. The current document, which has been finalised by the Secretariat, reflects the outcomes of the final discussions and deliberations on the alternative regulatory approaches proposed and incorporates two possible instruments. A Bilateral Protocol which focuses on the liberalization of traffic rights for air cargo services, ancillary services and other specific air cargo transportation issues that can be dealt with separately under existing bilateral air service agreements. And a possible Multilateral Agreement to facilitate early liberalization of international air cargo services without compromising essential safety and security aspects of civil aviation. Such an agreement could provide a means for liberalization of existing market restrictions on a multilateral basis.[11]

## 3. The World Trade Organization and the GATS

### *3.1 Implication for International Air Transport Policy*

Generally speaking, the WTO is the only international organization dealing with the global rules of trade between nations. The Organization came into being in 1995[12] as the successor to the *General Agreement on Tariffs and Trade* (GATT). At its core are the WTO agreements, the legal ground-rules for international commerce and for trade policy. The agreements have three main objectives: to help trade flow as freely as possible, to achieve further liberalization gradually through negotiation, and to set up an impartial means of settling disputes. By lowering trade barriers, consumers and producers can enjoy secure supplies and greater choice with respect to products and services. WTO's rules – the agreements – are the result of negotiations between the members. The current set was the outcome of the 1986–1994 Uruguay Round negotiations, which include a major revision of the original GATT and a *General Agreement on Trade in Services* (GATS),[13] which brought services within a multilateral framework of principles and rules similar to that covering trade in goods under GATT. The GATS contains a number of

---

[11] *Ibid.*

[12] The Uruguay Round of Multilateral Trade Negotiations under the GATT was launched in Punta del Este, Uruguay, in September 1986, and lasted till 1994, when the *Marrakesh Agreement Establishing the World Trade Organization*, (33 I.L.M. 1144 (entered into force 1 January 1995)) was signed [hereinafter the *Marrakesh* Agreement].

[13] *Ibid.*, Annex 1B.

annexes that cover special situations of individual service sectors, including the Annex on Transport services.[14]

Air transport services were clearly included in the GATS because the parties negotiating that agreement did not want to exclude any service sectors from progressive liberalization; however, they were not really eager to do so. Thus, the annex coverage applies to only three air transport services: aircraft repair and maintenance, the selling and marketing of air transport, and computer reservation systems.[15] It specifically excludes traffic rights[16] and all the services directly related to the exercise thereof. The reluctance of the aviation community to subject the sector to the GATS process stems from the basic GATS principles of most favored nation (MFN), national treatment, and transparency, which do not correspond to the present (bilateral) rules and policies related to international air

---

[14] In fact, GATS is made up of four components: the framework agreement, containing basic obligations applying to all Member States; national schedules of commitments made by States, specifying the modes or mode of delivery and any conditions on the market areas covered; national lists of exemptions from the obligations of Article II, dealing with Most Favored Nation (MFN) treatment; and a number of annexes. Three features of the GATS are particularly important: the GATS defines a process aimed at the progressive removal of barriers to trade in services; the aim is to cover all tradable services in all sectors; and the balance of benefits for a country is measured in relation to trade in all goods and services and not in just any one sector.

[15] V. Rodriguez Serrano, 'Trade in Air Transport Services: Liberalizing Hard Rights' (1999) 24 Air & Space L. 199 at 203-204.

[16] Article 6(d) of the Annex defines 'traffic rights' as follows: '"Traffic rights" means the right for scheduled and non-scheduled services to operate and/or to carry passengers, cargo and mail for remuneration or hire from, to, within, or over the territory of a Member, including points to be served, routes to be operated, types of traffic to be carried, capacity to be provided, tariffs to be charged and their conditions, and criteria for designation of airlines, including such criteria as number, ownership, and control'. These rights are also called 'hard rights', 'they have direct economic value because they give access to routes, hence to markets'; as opposed to 'soft rights', that 'are rights granted to the airlines of the signatory States of a bilateral or multilateral agreement in the territory of another signatory that do not, by themselves, have direct economic value. Soft rights typically include ground handling at airports, services, maintenance, and more recently, computer reservation systems or code sharing. They are accessory to the exercise of hard rights', see B.M.J. Swinnen, 'An Opportunity for Transatlantic Civil Aviation: from Open Skies to Open Markets?' (1997) 63 J. Air L. & Com. 249 at 250.

transport.[17] Moreover, other particularities of GATS make the Agreement difficult to apply to air transport services.[18]

The concern here is whether the air transport sector, and more specifically the airline ownership issue, can be liberalized within the GATS framework, as it provides a multilateral structure that allows for the globalization of industries, or whether the differences of perspective between the main GATS principles and the air transport principles are too great to realize liberalization in this structure. On the one hand, the air transport sector remains a particular industry, controlled by its own principles, and divergent from those of any other business sectors. Thus, the air transport liberalization process is led largely by the industry itself, and the current system of exchanging air transport rights reciprocally, through regional and plurilateral agreements among like-minded countries, has been slow but reasonably successful so far and may continue to work well. In addition, the main feature of air transport, which is traffic rights, has been specifically excluded from the Annex, and this includes the ownership and control issue. Therefore, if it is decided that the liberalization process should be achieved within the GATS framework, it must bring something more than the current system,[19] and it must take into account, at the same time, the industry particularities by modifying the Annex. On the other hand, air transport is a very strong business, so ignoring WTO and the GATS in the liberalization process would be unrealistic. In our view, the air transport sector should be completely included in the GATS framework for two main reasons.

---

[17] For more details about the application's concerns of these three GATS principles to air transport services, see R. Janda, 'Passing the Torch: Why ICAO Should Leave Economic Regulation of International Air Transport to the WTO' (1995) 21 Air & Space L. 409 at 417-418; R. Katz, 'The Great GATS' *Airline Bus.* (September, 1995) 81; G.L.H. Goo, 'Deregulation and Liberalization of Air Transport in the Pacific Rim: Are They Ready for America's "Open Skies?"' (1996) 18 U. Haw. L. Rev. 541 at 566-567; G.L.H. Goo, 'Deregulation and Liberalization of Air Transport in the Pacific Rim: Are They Ready for America's "Open Skies?"' (1996) 18 U. Haw. L. Rev. 541 at 468-469; C. Thornton & C. Lyle, 'Freedom's Paths' *Airline Bus.* (March 2000) at 74.

[18] For instance, 'the potential conflict (and consequently a need for clarification and harmonization) where the same activity is subject to both the GATS and other arrangements which are not included in bilateral or multilateral agreements concluded before 1 January 1995', see ICAO, Working Paper (*Report by the Council on Trade in Services*), A31-WP/23, EC/3 (7 April 1995) at 3; in addition, 'the GATS conflicts with some of the basic principles of the Chicago Convention', see Lehner, *ibid.*

[19] As Mr. R. Loughlin, from the US DOT, stated, in 2002 in a ICAO meeting:

> any proposition to *add* or *substitute* another system for exchanging and enforcing air transport rights must meet two tests: first, it must not put at risk the very significant gains already made, and the current operating freedom enjoyed, under the reciprocal system; second, it should be convincingly shown to promise improvements that could not be obtained under the existing and evolving bilateral/plurilateral/regional regimes.

R. Loughlin, 'The current GATS Round in a Historical Perspective' (Dialogue on Trade in Aviation-Related Services, ICAO, Montreal, 12 June 2001) [unpublished].

First, most of the economic sectors are now regulated by the 'WTO agreements'[20] and the air transport sector should not remain on the fringe of the world trade system.[21] One of the reasons is the current inter-dependence of the world and of all the economic sectors.[22] If the 'international economic system of today is the notion of 'liberal trade', meaning the goal to minimize the amount of interference of governments in trade flows that cross national borders,'[23] then, given this inter-dependence, that goal can be reached only if the whole economy is controlled by the same liberalized system. The second reason why the air transport sector should be totally regulated by the GATS framework is the efficiency of the whole WTO system, which benefits governments, consumers, and the industries themselves.[24] To give only three examples of the important Uruguay Round achievements, it provides a sophisticated dispute-settlement process for all portions of the 'WTO agreements', and provides a legal text (rather than just a customary practice) to carry out its procedure. These new procedures include measures to avoid 'blocking', in other words the impossibility of reaching a decision, which occurred under previous consensus decision-making rules. The agreement also provides for a new 'appellate procedure', which will be substituted for some of the procedures that were vulnerable to blocking. This is the only worldwide dispute-settlement mechanism that exists in the world, and is therefore necessary to the effective implementation of international economic rules. Indeed, the WTO is the most powerful organization with sanction and repression powers.[25] In addition, the Uruguay Round achieved an agreement on safeguards and escape-clause measures, and provides domestic adjustment assistance policies whereby '[e]ach of the major industrialized nations has adopted its own policy approaches to the challenge of economic adjustment, including adjustment to trade liberalization.'[26] As a third reason, the GATS, together with the WTO Final Act, offer developing countries certain positive guarantees regarding liberalization, as it 'is designed to allow for

---

[20] The 'WTO agreements' are annexes of the *Marrakesh Agreement* (it includes the *Multilateral Agreements on Trade in Goods* (Annex 1A), the *General Agreement on Trade in Services* (Annex 1B), the *Agreement on Trade-Related Aspects of Intellectual Property Rights* (Annex 1C), the *Understanding on Rules and Procedure Governing the Settlement of Disputes* (Annex 2), the *Trade Policy Review Mechanism* (Annex 3), the *Plurilateral Trade Agreements* (Annex 4)); ministerial decisions and declarations; understanding on commitments in financial services. For more details about each agreements, see M.J. Trebilcock & R. Howse, *Regulation of International Trade*, 2nd ed. (London: Routledge, 1999); see also T. Flory, *L'Organisation Mondiale du Commerce – Droit institutionnel et substantiel* (Bruxelles : Etablissements Emile Bruylant, 1999).

[21] '[W]hatever the special features of the air transport industry, the GATS excludes no service sector and the Annex on Air Transport Services is formulated so as to apply, in principle, to all aspects of the industry', see Janda, *supra* note 17 at 419.

[22] J.H. Jackson, *The World Trading System – Law and Policy of International Economic Relations*, 2nd ed. (Cambridge: the MIT Press, 1997) at 6.

[23] *Ibid.* at 11.

[24] Janda, *supra* note 17 at 409.

[25] About the dispute-settlement mechanism, see Jackson, *supra* note 22 at 107; see also Trebilcock & Howse, *supra* note 20 at 58.

[26] Trebilcock & Howse, *ibid.* at 239.

the coexistence of various domestic regulatory approaches, including departures from the MFN principle, while seeking progressive liberalization.'[27]

Considering all the above, the GATS framework appears to be an unavoidable and a necessary tool to achieve air transport liberalization, including the liberalization of hard rights. In order to complete this liberalization, the services covered by the air transport Annex need to be expanded in order to remove the specific hurdles to trade that exist in air transport.

## 3.2 Implication of the Air Transport Liberalization Process: Extension of the Annex on Air Transport Services

Achieving air transport liberalization imposes the extension of the Annex on air transport services by including traffic rights (which essentially encompass the seven air transport freedoms) and all the services that are directly related to their exercise. According to Article 6(d) of the Annex,[28] it involves the capacity, tariff, designation of airlines, and the ownership and control issues. The extension of the Annex has been foreseen in the text itself, as Article 5 states that the scope of the Annex shall be reviewed every five years, which aimed at achieving a higher degree of liberalization.[29] In the preparation of such a review, extensive discussions have taken place in the past at various levels regarding the activities that could potentially be included in the Annex. Thus, the round of negotiations for review was launched by the third WTO Ministerial Conference held in Seattle from 30 November to 3 December 1999. This conference was suspended on the last day without any decision being adopted. Today, the negotiations continue;[30] however, it is still unlikely that 'hard rights' be included in the Annex. The reticence of many States to include them and the important principle of decision by consensus among WTO members are two obstacles that slow down the liberalization of market access and of airline ownership and control in the GATS framework.

In order to reach decisions in the rounds of negotiations, the Annex expansion will probably be incremental and will come in two stages. The first stage

---

[27] Janda, *supra* note 17 at 428.

[28] Definition of 'traffic rights', *supra* note16.

[29] Article 5 of the Air Transport Annex goes on to specify that:

> The Council for Trade in Services shall review periodically, and at least every five years, developments in the air transport sector and the operation of this Annex with a view to considering the possible further application of the Agreement in this sector.

[30] The last Ministerial Conference met in Doha, Qatar, in November 2001. Before, the WTO has also held meetings in Geneva in 2000 to explore and exchange views on the subjects that could be included within the GATS agreement. '[The] last meeting considered, *inter alia*, the addition of a specific annex for tourism services to the GATS, which would include air transport services as one of the activities related to tourism. The sectorial classification attached to that annex has shown that the relevant transport services include scheduled and non-scheduled air transport services for passengers, airport ground handling, cargo, repair service, fuel, etc, see A.J. Al Dawoodi, 'The impact of the GATS on air transport from the general perspective of developing countries' (Dialogue on Trade in Aviation-Related Services, ICAO, Montreal, 12 June 2001) [unpublished].

is to clarify and strengthen the current Annex and to include new soft rights. Basically, it means obtaining a more equivalent treatment on the basis of reciprocity, as almost half of the States took 'MFN exemptions'[31] with respect to marketing and selling of CRS.[32] Therefore, existing exemptions should be the subject of new negotiations. Clarification is also needed with respect to the definition of services given in the Annex.[33] Furthermore, from the WTO meetings, it is clear that some States would like to bring a number of services under the GATS umbrella, including cargo, express mail, non-scheduled services for passengers, and ground handling services (such as fuelling, cleaning, air catering).[34] This first undertaking is certainly the easiest step towards liberalization, even though a few rounds of negotiations will still be necessary. The second stage of the Annex expansion is its extension to hard rights. According to Article 6(d) of the Annex, the ownership and control issue is part of this stage. Despite the language of the Annex itself which would make it easy to include traffic rights,[35] it will certainly take many years before reaching a consensus on these issues. Indeed, apart from two sectors that have already received support for future inclusion under the scope of the GATS (e.g. air cargo services and non-scheduled flights),[36] most of the States are not yet ready to proceed with such a revolutionary shift from the current bilateral system and the MFN principle is their first concern.[37] Various solutions have been proposed to solve this problem. One solution would alter the nature of the MFN obligation to make it a 'conditional MFN under which countries that mutually agree to accept higher levels of obligation should not be required to extend the same treatment to countries which were unwilling to do so'.[38] Some authors have proposed a specialized application of the MFN clause based upon mirror reciprocity. It would mean that 'every WTO member would offer to every other member the equivalent of the most favorable bilateral arrangement into which it is currently prepared to enter on the basis of mirror reciprocity.'[39] Through such an application, there could be no situations where countries could take advantage of the MFN clause without offering reciprocal benefits. Thus, developed countries as well as developing countries would be protected, as no one would be

---

[31] A State which takes a 'MFN exemption' means that it is not obliged to offer MFN treatment in the sector.

[32] Katz, *supra* note 17.

[33] Rodriguez Serrano, *supra* note 15 at 209-210.

[34] *Ibid.* at 208-209; Loughlin, *supra* note 19.

[35] 'This argument focuses on the inclusive character of GATS provisions, and the exclusion of traffic rights as exceptions, for "[i]n an inclusive agreement with exceptions it is always easier to remove the exception than to extend the scope of the agreement"', Rodriguez Serrano, *supra* note 15 at 211.

[36] *Ibid.* at 211-212; ICAO, Working Paper (*Report of the Council on Trade in Services*) No. A33-WP/7 (5 June 2001) at 4 [hereinafter ICAO Working Paper NO. A33-WP/7].

[37] As far back as the 1940s, some attempts were made to utilize the MFN clause with bilaterals, see Little, *supra* note V. Little, 'Control of International Air Transport' (1949) III International Organization 29 at 37.

[38] e.g. Lehner, *supra* note 17 at 470.

[39] Janda, *supra* note 17 at 423-424; Rodriguez Serrano, *supra* note 15 at 213-214.

committed by any obligation if it finds that this obligation could not be adapted to its own environment.

These proposals could alleviate the State concern regarding the liberalization of difficult air transport issues, such as market access and ownership and control. However, further detailed studies need to be conducted and no decision on these issues is expected for this round of negotiations. The very complex legal procedure in WTO is not the only reason why a consensus cannot be obtained. Political conflict is one of the principal reasons. Another main reason is probably that very few States have sufficient confidence in the survival of their airlines in case of multilateral (GATS) liberalization. However, the liberalization process is not closed. Market access and foreign investments will be liberalized progressively through bilateral, regional, and plurilateral agreements. In parallel, ICAO continues to play its role in the liberalization process, by pursuing multilateral discussions on these specific topics. All these initiatives will surely help governments to find a consensus within a few years through the GATS framework.

## 4. The International Civil Aviation Organization

### *4.1 The Prominent Role of ICAO in the Air Transport Liberalization Process*

ICAO is the worldwide intergovernmental organization created by the *Chicago Convention* of 1944,[40] for the promotion of the safe and orderly development of international civil aviation throughout the world. As a specialized agency of the United Nations, it sets international standards and regulations that are necessary for safe, regular, efficient and economical air transport.[41] It also serves as a medium for co-operation in all fields of civil aviation among 187 Contracting States (as of 25 February 2001).[42] Policy is developed in multilateral meetings, and a significant volume of preliminary research and analysis, which support the meetings of policy development bodies, is achieved by the ICAO Secretariat. Moreover, ICAO provides Contracting States with various published statements of its policy on international air transport regulatory matters, as developed or endorsed by the Assembly or the Council, as well as guidance materials and information developed by ICAO bodies or the Secretariat.[43] With respect to air transport liberalization, ICAO has the mandate, experience, and expertise in a wide range of air transport matters – technical, economic and legal. Abeyratne stresses the fact that:

> [M]ultilateralism in the form of a broad-based consensus on principles and guidance to States in the conduct of their air transport activities has enjoyed renewed interest

---

[40] *Convention on International Civil Aviation*, 7 December 1944, 15 U.N.T.S. 295, ICAO Doc. 7300/6, Article 43 [hereinafter the *Chicago Convention*].
[41] *Ibid.* Article 37 and Article 38.
[42] *Ibid.* Article 44, Article 54, and Article 55.
[43] See ICAO Doc. AT/122 (9 October 2001) at 2.5 [hereinafter ICAO Doc. AT/122].

in ICAO in recent years. While seeking to progressively develop positions and guidance to assist States in their regulatory/economic activities, ICAO recognizes the sovereignty of States in pursuing their own national air transport policies and objectives. ICAO's role in this sphere is therefore merely **consultative** and **recommendatory** without being incompatible with liberalization in this sector.[44]

It is important to recall briefly ICAO's role because it has been very frequently misunderstood in recent years. Indeed, the Organization has often faced criticism on different points. It has been reproached, *inter alia*, on the ground that the *Chicago Convention* requires a system of air transport regulation based only on bilateral agreements or restrictions, which makes it incompatible with air transport globalization and liberalization trends. However, it can be argued that the key provisions of the Convention are mainly the recognition that every State has complete and exclusive sovereignty over the airspace of its territory (Article 1) and that no scheduled international air service may be operated over or into the territory of a contracting State except with a special permission or authorization of that State (Article 6). The terms of special permission or authorization, based on reciprocity, can be strict or liberal, negotiated bilaterally or multilaterally.[45] Thus, despite certain necessary amendments,[46] the *Chicago Convention* is adaptable to the current environment. Another flaw of the Convention which has been emphasized by the international community is the inadequacy of its dispute settlement mechanism. The dispute settlement Articles of the Convention (Article 84-88) have been invoked in five instances to date, three of which concerned disputes over sovereign airspace infringements; all three were settled through diplomatic negotiations, as the Council never issued a decision on the merits of the case.[47] This argument, however, does not make ICAO ineffective for a role in the international air transport liberalization process and the assistance of States in their economic activities. In addition, the bilateral and regional agreements themselves usually have their own dispute settlement system. Given their respective roles, WTO's framework is certainly better equipped to be in charge of an international judicial system than ICAO's structure. Finally, ICAO has also been criticized for being unable to make binding legal decisions or to reach concrete results at their meetings or international conferences thus far.[48] However, even though the outcomes have not always been as expected, ICAO represents a discussion forum

---

[44] R.I.R. Abeyratne, 'Emergent Trends in Aviation Competition Laws in Europe and in North America' (2000) 23 World Competition. R. 141 at 160 [emphasis added][hereinafter Abeyratne 'Aviation Competition Laws'].

[45] Thornton, Lyle, *supra* note 17 at 74.

[46] M. Milde, 'The Chicago Convention – Are Major Amendments Necessary or Desirable 50 Years Later' (1994) 19:1 A.A.S.L. 401 at 414-447.

[47] The three cases were: India versus Pakistan 1952/1953, Pakistan versus India 1971/1976, Cuba versus United-States 1996/1998.

[48] So far, four ICAO air transport conferences were held in 1977, 1980, 1985, and 1994 respectively. Apart from the 1994 conference, the first three 'conferences were by no means comprehensive in their deliberations and had only addressed specific issues at each conference', see Abeyratne *Competition in Commercial Aviation, supra* note 1 at 814.

rather than a place of decision-making. By providing a forum where States exchange their ideas about the economic regulation of international air transportation, ICAO provides States with a better understanding of the main air transport issues.[49] According to the role defined above, the Organization has performed its functions properly so far, and still has an important role to play in further assisting States to understand developments and prospects of national and international civil aviation.

ICAO is competent and needs to take an active role in developing future regulatory arrangements in two areas. First, ICAO has to retain its leadership role in regulating technical matters, such as safety, security and the environment. In addition to specific provisions in the *Chicago Convention* related to the prevention of civil aviation incidents,[50] ICAO has been very active in regulating these big issues. For instance, one of the main achievements was the creation of Universal Safety Oversight Audit Programme in January 1999.[51] The current air transport liberalization process cannot be achieved properly, whether this process is led by the industry itself or by an organization, without a strict and clear regulation in the field of safety and security. The ownership and control issue is a good example of the need for such a regulation.[52] Therefore, ICAO should be recognized and acknowledged as the worldwide auditor of safety and security standards for international civil aviation. There is no organization better equipped to deal with these issues. Second, ICAO has an economic vocation. The *Chicago Convention* mandated to ICAO the development of principles and techniques regarding the economic air transport development,[53] and it has become more and more involved in aviation trade-related services, especially since the 1994 WorldWide Air Transport Conference.[54] After the Conference, the ICAO Council was assigned the study of certain issues related to future regulatory arrangements for international air transport to the Air Transport Regulation Panel (ATRP). In 1997, the ATRP adopted certain recommendations regarding market access for carriers.[55] Moreover,

---

[49] *Ibid.* at 803.

[50] The *Chicago Convention*, Article 9-17, Article 24-26, *supra* note 40.

[51] About ICAO's actions regarding the civil aviation safety, see Part 2, Chap.3, para. 3, above.

[52] Indeed, in the event of the liberalization of foreign investment, the main concern is the risk of flags of convenience, see *ibid.*

[53] The *Chicago Convention*, Preamble, article 44, *supra* note 40.

[54] Unlike the first three ICAO air transport Conferences, the 1994 Conference started to take into account commercial aspects of air transport, such as market access, ownership and control, capacity, and pricing issues. This Conference has been successful in that it has made the international community of States more conscious about the need to liberalize. For more details about the economic role of ICAO since 1994, see *e.g.* ICAO Working Paper No. AT Conf/4 – WP 7, *supra* note 58 at 4; ICAO, Working Paper (*World-Wide Air Transport Conference on International Air Transport Regulation: Present & Future*) No. AT Conf/4 – WP 10 (19 April 1994) at 4-6; Abeyratne *Competition in Commercial Aviation, supra* note 1 at 809-825.

[55] Thus, the Council of ICAO has issued a series of recommendations as a guide to the Organization's Contracting States in adjusting to an increasingly competitive airline

in July 1998, the WTO Council for Trade in Services decided to confer observer status upon ICAO, which was thereby permitted to attend WTO meetings on an *ad hoc* basis.[56] Presently, ICAO works continuously on the key regulatory issues that would need to be resolved to enable the international community to move towards further globalization.[57]

Accordingly, ICAO is certainly a 'nimble, networked agency ... that can develop better modes of consultation and engagement with global civil society.'[58] As such, the Organization has an essential place in the air transport liberalization process. Its economic role should be more acknowledged as complementary to the role played by the WTO. Furthermore, it 'has clearly a vital role in respect of safety, security, environment, transit, etc., without which full liberalization of market access would be impossible ... [therefore] ICAO in many ways could facilitate the whole process.'[59]

### 4.2 ICAO Activities with Respect to the Liberalization of Airline Ownership and Control

As stressed by Dr. Assad Kotaite, President of the ICAO Council, the relaxation of ownership legislation remains a major policy goal for ICAO.[60] In fact, ICAO started seriously discussing the airline ownership and control issue in 1992, when the Worldwide Air Transport Colloquium convened by the Council identified the problems connected with multinational ownership of airlines that would deserve future actions.[61] In 1994, ICAO was at the forefront of guidance in this area, as the 'airline substantial ownership and effective control' principle was one of the main

---

environment while fostering fair competition. The recommendations are the conclusion of several years of work on the broad issue of the economic regulation of international air transport (among them, may be noted the proposals to broaden the ownership and control criteria, to create an aviation industry-focused dispute settlement mechanism for use in the liberalizing environment, and to elaborate a series of model clauses for use in bilateral or multilateral air services agreements), see Abeyratne *Competition in Commercial Aviation*, *ibid.* at 826-829; Rodriguez Serrano, *supra* note 15 at 206-208; D., Hughes, 'ICAO delegates Shun US Free-Market Stance' *Aviation Wk. & Space Tech.* (2 January 1995) 37.

[56] Rodriguez Serrano, *ibid.* at 205-206.

[57] ICAO Working Paper No. A33-WP/7, *supra* note 36 at 2; 'Progressive Liberalization Actively Supported by ICAO, Council President Tells IATA Annual General Meeting' (June 2001) 56 ICAO J. 30 [hereinafter ICAO J. Doc. 2001].

[58] R. Janda, 'ICAO as a Trustee for Global Public Goods in Air Transport' (Dialogue on Trade in Aviation-Related Services, ICAO, Montreal, 12 June 2001) [unpublished].

[59] F. Sørensen, 'Market Access Liberalization of Air Transport' (Dialogue on Trade in Aviation-Related Services, ICAO, Montreal, 12 June 2001) [unpublished].

[60] WTO, *Note on Developments in the Air Transport sector Since the Conclusion of the Uruguay Round, Part Five*. WTO Doc. S/C/W/163/Add.4 (2001) 7 [hereinafter WTO doc.1].

[61] J.S. Gertler, 'Nationality of Airlines: Is It a Janus with Two (or More) Faces?' (1994) 19:1 A.A.S.L. 211 at 222-225.

subjects addressed at the Fourth Worldwide Air Transport Conference.[62] At that time, within the framework of the ICAO Conference, ICAO bodies, national and international organizations, and States analyzed the issue and drafted comments on different topics, such as the necessary change of the national restrictions, the way to broaden the ownership and control criteria, the right of establishment, and the risks at stake in case of a lifting of the restrictions.[63] Then, in 1997, after careful consideration of safety risks and other problem involved, the ICAO ATRP recommended the broadening of ownership and control criteria used in bilateral agreements, by proposing the criteria of 'the principal place of business' and 'a strong link'.[64] Meetings organized by ECAC on the questions of ownership and control followed in 1998 and 1999. Seventeen States and seven organizations exchanged their points of view, specifically on 'ownership and control in the bilateral agreements' and on 'the development of a model clause on a strong link between the air carrier and the designating State', in light of the 1997 ICAO recommendations.[65]

In the year 2001, the ownership liberalization issue was again high on the agenda. The ICAO Secretariat initiated a comprehensive study and, as a first step, launched a survey on States' relevant policies and practices. In May of that year, a questionnaire[66] was sent to all ICAO Member States to gather information and views from States. A year later, the Secretariat presented to the Council the results of the survey and its intentions regarding further work on the subject.[67] According

---

[62] R.I.R. Abeyratne, *Emergent Commercial Trends and Aviation Safety* (Aldershot: Ashgate, 1999) at 23 [hereinafter Abeyratne *Emergent Commercial Trends*].

[63] See e.g. ICAO, Working Paper (*World-Wide Air Transport Conference on International Air Transport Regulation: Present & Future*) No. AT Conf/4 – WP 8 (20 April 1994); ICAO, Working Paper (*World-Wide Air Transport Conference on International Air Transport Regulation: Present & Future*) No. AT Conf/4 – WP 9 (21 April 1994); ICAO, Working Paper (*World-Wide Air Transport Conference on International Air Transport Regulation: Present & Future*) No. AT Conf/4 – WP 18 (20 July 1994); ICAO, Working Paper (*World-Wide Air Transport Conference on International Air Transport Regulation: Present & Future*) No. AT Conf/4 – WP 30 (12 August 1994); ICAO, Working Paper (*World-Wide Air Transport Conference on International Air Transport Regulation: Present & Future*) No. AT Conf/4 – WP 47 (23 August 1994); ICAO, Working Paper (*World-Wide Air Transport Conference on International Air Transport Regulation: Present & Future*) No. AT Conf/4 – WP 52 (8 September 1994); ICAO, Working Paper (*World-Wide Air Transport Conference on International Air Transport Regulation: Present & Future*) No. AT Conf/4 – WP 68 (20 October 1994).

[64] ICAO, *Broadening of Ownership and Control Criteria*, ICAO Doc. No. ATRP/WP/5 (14 November 1996); ICAO Doc. 9587, *supra* note 108 at 2-1.

[65] ECAC Doc. 1998, *supra* note 108; ECAC, *Report on Task Force on Ownership and Control Issues, Second Meeting*, OWNCO/2 (26 February 1999).

[66] ICAO, *Questionnaire on State's Policies and Practices Concerning Air Carrier Ownership and Control*, Attachment to State letter SC 5/2-01/50, online: ICAO http://www.icao.int/cgi/goto_atb.pl?icao/en/atb/ecp/S150-survey.htm;ecp (date accessed: 2 January 2001).

[67] ICAO, Working Paper (*Result of the Survey of States' policies and practices concerning air carrier ownership and control*) No. AT-WP/1933 (2 April 2002).

to this working paper, it is reported that, by 28 February 2002, 54 replies had been received.[68] The results of the survey lead to three main conclusions. One of the important findings of the survey relates to the rationale why States regulate air carrier ownership and control. Today, national development/economic interests is considered by States as the most important justification of the remaining restrictions, whereas the aviation safety and national security reasons are on the fifth and seventh position out of eight reasons, rated according to their degree of importance. It should also be stressed that conformity with international agreements is the second most important reasons for States to keep the restrictions.[69] This remark leaves the door opened for ICAO to reform the regulatory arrangements at the international level (primarily relevant bilateral provisions). The second main concluding remark of the survey is, unlike the 'substantial ownership' notion, the remaining ambiguity prevailing on the 'effective control' definition. Indeed, only 23 States indicated that they have a specific provision which determines what constitutes 'effective control' of an air carrier in their national legal regime. Thus, the diversity of interpretations, depending on the national interest protection, of what constitutes control of an air carrier among States explain the difficulty for the Community of States to apply a uniform and objective treatment to all air carriers.[70] Finally, the third remark concerns the evolution of the traditional ownership and control clause towards broadened criteria. In this regard, only 10 per cent of all the bilateral agreements have provisions that deviate from the 'traditional criterion'[71] of ownership and control, however, many States would be prepared to accept alternative criteria, such as the creation of a multilateral carrier by an intergovernmental agreement (i.e. SAS and Gulf Air) or the incorporation and the principal place of business or permanent residence of the airline in the designating State. This relatively significant acceptance to broaden the traditional criteria call also for a new definition of the regulatory arrangements.[72] Thus, proposals for a reforming international framework, as well as for alternative regulatory arrangements for airline designation and authorization,[73] were presented to the Air Transport Regulation Panel (ATRP) at its tenth meeting in May 2002.[74] Furthermore, in

---

[68] Although this number of respondents is relatively low, ICAO has considered that the respondents provide a good representation in geographical regions as well as a cross-section of States, and therefore, 'the Survey has met its objective in providing a substantive basis for the study', see *ibid.*, at 4

[69] *Ibid.*

[70] *Ibid.*

[71] That is, the designated airline must be substantially owned and effectively controlled by the designating Party or its nationals.

[72] ICAO doc., *supra* note 64 at 4.

[73] 'Efforts should be made to ensure that any regulatory arrangements developed be widely acceptable, practical in application (without disrupting the existing regulatory process) and address the needs and concerns of States, particularly those of the developing countries', *ibid.* at 5.

[74] ICAO, *Report of the Air Transport Regulation Panel Working Group on Air Carrier Ownership and Control*, ICAO doc. (September 2002), online: ICAO

2001, Egypt proposed the application of a principal place of business test under certain conditions with the objective of promoting and consolidating the air transport regulatory regime and achieving a higher degree of flexibility as to the acceptance of the designation of foreign airlines.[75] And in September 2001, the United Kingdom called for further urgent liberalization measures, especially due to the great pressure caused by the 11 September attacks affecting the aviation industry. In addition, the UK stressed the important role of ICAO by stating that it is 'incumbent upon ICAO to display leadership in helping to sustain the viability of the industry through these difficult times by urging States to use the flexibility already available to them in interpreting their bilateral obligations.'[76] In October 2001, the ICAO Economic Commission has underlined that foreign investments in airlines 'might be an overriding concern in the near future' and that the Secretariat 'has already commenced work on studying this subject further in preparation for AT Conf/5.'[77]

The next important ICAO meeting devoted to the air transport liberalization is the Fifth ICAO Air Transport Conference on 'Challenges and Opportunities of Liberalization', which is set to take place from 24 to 29 March 2003. The objective of the Conference is 'to develop a framework for the progressive liberalization of international air transport, with safeguards to ensure fair competition, safety and security, [and including] measures to ensure the effective and sustained participation of developing countries.'[78] As mentioned earlier, ICAO's aim is not to take any decisions, but rather to provide a global forum for ICAO Member States and other concerned parties to examine issues and policy options in the field of air transport regulation and promote a better understanding of the concept and the impact of full liberalization. The conference will mainly focus on the key regulatory issues associated with liberalization, the review of a template air services agreement, and the adoption of a declaration of global principles for international air transport.[79] Air carrier ownership and control is one of the key regulatory issues that will be examined at the 2003 Conference.[80] The results of the 2001 ownership and control ICAO survey, which has revealed in particular that a broader perspective of national interest including economic development and trade

---

http://www.icao.int/atrp (date accessed: 1 October 2002).

[75] ICAO, Working Paper (*Substantial Ownership and Effective Control of Designated Airlines*) No. A33-WP/96 (17 August 2001).

[76] ICAO, Working Paper (*Substantial Ownership and Effective Control over Designated Airlines*) No. A33-WP/181 (25 September 2001) [hereinafter ICAO Working Paper No. A33-WP/181].

[77] ICAO, Working Paper (*Economic Commission – Draft Text for the Report on Agenda Item 26*) No. A33-WP/262 (2 October 2001).

[78] 'Progressive Liberalization Actively Supported by ICAO, Council President Tells IATA Annual General Meeting' (June 2001) 56 ICAO J. 30 [hereinafter ICAO J. Doc. 2001].

[79] *Ibid.*

[80] The issues that will be examined are air carrier ownership and control, market access, product distribution, fair competition and safeguards, conditions of carriage and consumer protection, extra-territorial friction, dispute resolution, and registration and transparency of air services agreements.

has become an overriding factor, will serve as a basis of discussion. In addition, in its preparatory papers, ICAO has discussed and proposed possible approaches to facilitating liberalization with respect to the ownership and control issue.[81] *Inter alia*, ICAO has proposed to the international Community an article providing a practical alternative for States to liberalize conditions regarding airline designation and authorization in their air services agreements, which effectively address the needs and concerns of States.[82] Moreover, ICAO stressed that 'while it will be up to each State to choose its liberalization approach and direction based on national interest, the use of the proposed arrangement could be a catalyst for broader liberalization.'[83] Furthermore, ICAO is currently preparing the Conference in collaboration with the WTO and the OECD so that all the possible options will be presented and analyzed at the Conference in order to guide States in updating the 'designation clause' in their bilateral, regional, and plurilateral agreements in the context of the air transport liberalization. ICAO, with the 2003 Conference, does not intend to go beyond the limits of what is realistically feasible.[84] This means that, in an environment substantially more open to liberalization than at the 1994 worldwide air transport conference, ICAO and all the parties present at the Conference will 'simply' address the global principles and help States to move towards liberalization. In addition, the idea of drafting, in the long run, a multilateral agreement on air transport liberalization may be raised as well, but probably only from the perspective of a first multilateral approach.

## 5. Conclusion

'Cooperation' is the emphasis of this chapter. Cooperation is required and is indeed the norm between the international organizations involved in the air transport liberalization process. ICAO, WTO, and OECD do not compete with each other: they are complementary and not substitute organizations. The WTO should continue with expanding the existing scope of the GATS Air Transport Annex to hard rights, as the GATS is the proper framework for progressive multilateral liberalization. However, the WTO cannot lead this process alone. ICAO, as the universal expert in the field, must pursue its discussions and continue drafting international regulations in the technical and economic fields in order to ensure a viable liberalization of, mainly, market access and foreign investments. The several initiatives regarding air transport liberalization undertaken by ICAO these past years and the coming worldwide Conference (ATConf/5) demonstrate the leading

---

[81] ICAO, Working Paper (*Liberalizing Air Carrier Ownership and Control*) No. ATConf/5 WP/7 (21 October 2002).

[82] *Ibid.*, at 5.

[83] *Ibid.*, at 9.

[84] The ICAO ownership and control study 'has proven to be unrealistic and unproductive for ICAO to attempt to achieve global consensus on how States should adjust their national legislation, rules or policies governing their own national carriers', see ICAO doc., *supra* note 64 at 4.

role of the Organization in facilitating liberalization. Finally, the OECD is the indispensable organization that guides the WTO, ICAO, and the States in pursuing their roles by identifying principal problems and by proposing strategic solutions. Facilitation of a State's negotiations is the main role and responsibility of these organizations. However, Members of the WTO, ICAO, and OECD are States, so the organizations cannot undertake any action without the good will of individual States to cooperate with one another. Effort to harmonize national regulations is urgently needed with respect to safety, security and fair competition in order to present undesirable consequences following liberalization of foreign investment. Furthermore, since harmonization and liberalization, in particular in the form of a multilateral agreement, will only come gradually, the negotiations should start as soon as possible between States.

# Conclusion

The time has come to offer the aviation industry a new regulatory framework, similar to those of other mature industrial sectors that benefit from a global and liberal market. Cross-border investments, through international M&As, are presently occurring in all sectors, and are particularly characteristic of service sectors that, as a result of regulatory reform, have seen the privatization and liberalization of trade and investment regimes and are now able to restructure more freely at both national and international levels. Unlike mergers in the past, mergers are currently motivated by the desire to consolidate the capacity to serve global markets and fully benefit from economies of scale. The airline industry should benefit from the same liberal environment; it is no longer an infant industry and should leave the restrictive bilateral system that has served the industry for 50 years, in order to gain more global strength. The main objective of this book was to identify the factors that still justify the imposition of national ownership restrictions on airlines; the present analysis demonstrated that none of the legal and economic reasons used justify these restrictions any longer. So far, a protectionist policy has prevailed in fact in several States, such as the US, probably to the detriment of the airline industry itself. Today, the only tenable reason for keeping national airline ownership restrictions is the national security concern; however, this issue can be resolved separately from the airline ownership and control issue. Accordingly, removal of foreign ownership restrictions should be a matter of priority for the respective national authorities worldwide, since it would play a paramount role in the future global consolidation of airlines and would be an important step towards the normalization of the industry.

The year 2001 has reminded the world of the fragility of the airline industry. In Europe, many air carriers faced serious financial problems: medium-sized carriers have been badly hit by an almost global malaise bordering on a recession, the increase of oil prices, demonstrating that they are not well-suited to withstanding economic shocks, while the larger European carriers have had difficulties competing with the highly consolidated American airline industry. As for the United States, the fall-out from the US global slowdown was already eroding the margins of the large carriers as transatlantic travel fell away while the world's largest economy entered into a recession.[1] The 11 September attack on the World Trade Center, which used civil aviation as a weapon of destruction, has exacerbated the global problems faced by an already fractured industry. Indeed, the attack has lead to a loss of confidence in the industry and in the span of three months, the downturn has claimed well-known national carriers such as Swissair

---

[1] P.Y. Dugua, 'Le ciel américain broie du noir' *Le Figaro* (23 June 2001) III.

and Sabena,[2] followed in 2002 by their giant US counterpart United Airlines.[3] Obviously, the aviation industry does not simply live in a short-term recession but rather in a long-term crisis.[4] Airlines are currently under unprecedented pressure. Even though some of the major airlines are starting to recover, their assets and stock market value are still very low. This whole situation clearly calls for consolidation[5] and, therefore, makes cross-border investments in the airline industry not only desirable, but a necessity in order to address the needs of the industry. It is now urgent that foreign investments be fostered.

Unfortunately, the development of cross-border investment faces two obstacles, which are also resulting from the crisis. First, an economic obstacle prevails. Indeed, as the entire industry is affected by the recession, no airline has any money to invest in other carriers. Furthermore, owing to the global recession of national economies, it would be risky for the investors outside the industry to invest their capital in such a fluctuating and unpredictable industry.[6] The second obstacle concerns national security. The 11 September terrorist attacks has resulted in a loss of confidence in the air transport industry. As airlines were misused as weapons of destruction, air travel became an unreliable and risky means of traveling. In the year ahead, the main challenge for the whole aviation industry is, therefore, to regain the confidence of the passengers. To this end, passengers should be able to rely on the airline industry, they need transparency, in other

---

[2] 'The head of the global airlines body IATA has forecast that global 2001 losses in the sector will be in the region of US$10 billion to $12 billion. Only during the Gulf War did losses approach this year's levels, and the bad news is that the problems facing the industry look set to continue well into 2002', see M. Glackin, 'Airline Consolidation is the only route to survival' *The Scotsman* (24 December 2001), online: The Scotsman http://209.185.240.250/cgibin/linkrd?Lang=EN&lah=7986709cb8b8d0bfa2360ae6558c6ed8 &lat=1010085559&hmaction=http%3a%2f%2fwww%2eairliners%2enet%2fnews%2fredire ct%2emain%3fid%3d30291 (date accessed: 24 December 2001).

[3] K. Alexander, 'United seeks bankruptcy protection' Washington Post (10 December 2002), online: Washington Post http://www.washingtonpost.com/wp-dyn/articles/A32793-2002Dec9.html (date accessed: 11 December 2002).

[4] IATA reports that 400,000 air transport workers have lost their job since 11 September 2001, and the airlines lost overall, $18 billion. 'In terms of 2002, [IATA] ha[s] figures for international scheduled traffic, which shows losses between $4 billion and £6 billion', see D. Gavlak, 'Airline Industry Suffered Massive Losses After September 11' voanews (11 September 2002), online: voanews http://www.voanews.com/PrintArticle.cfm?objectID=5E25A889-57BA-49EF-A89D2 (date accessed: 19 September 2002).

[5] *Ibid.*

[6] For instance, regarding the situation in Canada, '[Thomas Ross, a professor at the University of British Colombia] said raising the foreign cap would allow other airlines to come in and invest in Air Canada. But airlines around the world are short of cash because of a dramatic drop in bookings since the 11 September terrorist attacks. 'People inside the industry don't have much cash right now, and people outside the industry are probably looking at it and saying: "Is this really a place to put my money?"', see K. McArthur & S. Chase, 'Schwartz spurns Air Canada as Ottawa mulls ownership cap' *The Globe and Mail* (5 October 2001) B1.

words, they need to know who actually owns and controls the national airline. In this logic, it is not excluded that, for example, if BA were to be owned and controlled half by a Middle East airline, British people may not feel comfortable with their 'semi-national' airline. Accordingly, after being used as an instrument to protect national economies, the issue of airline ownership and control may have returned to the original concept of national identity. This is the case because passengers, concerned with safety and security, feel a need to deal with an airline with a national identity they can trust.

Does this mean that the liberalization process of the ownership and control restrictions has come to a standstill? Probably not, since the situation has not changed much in reality.

First, since the end of the Cold War, while the main reason for the reluctance of States *vis-à-vis* foreign investment liberalization had been economic, the original rationale has returned. Aviation safety and national security are the dominant preoccupation of virtually all States, and the US is certainly the most affected State. However this has always been the case: the US has always claimed the necessity of keeping national ownership restrictions to safeguard the CRAF Program. It is indeed legitimate to believe that political differences, even among the closest US allies, would have the potential to disrupt and undermine CRAF operations at any time, particularly where these operations are linked to controversial political issues, such as in the Middle East. Nevertheless, the international community of States should keep negotiating with the US in order to find a compromise among the diverse solutions mentioned above in the analysis of national security.[7] However, it will likely take a few years to convince the US to liberalize its ownership system. First, because politics interfere substantively in the lawmaking process – the independence of the legislative power is in fact very questionable in most of States – and, second, because the Administration, even though it wants to move forward, will not pick a fight with domestic labor or the military unless there is a big payoff. Nevertheless, due to the need for outside capital of US carriers, the US will likely be pushed to discuss changes in foreign ownership soon, and decide to proceed on a case-by-case basis. Considering the current financial crisis of the US airline industry, the need of foreign capital is obvious. Accordingly, once the US will accept more foreign capital in its industry, whether to a substantial extent or not, the principle of substantial ownership and effective control of airlines will be liberalized progressively through regional and plurilateral agreements, such as the *APEC Agreement* (in which the control criterion could also eventually be abolished).

Second, the liberalization process of ownership and control restrictions should not be slowed down because of the need for airlines to regain passenger confidence. Indeed, States are now less reluctant to liberalize their ownership policy since September 2001. Surprisingly, the replies to the ICAO questionnaire on ownership and control indicate that States, today, attach less importance to national ownership than they used to do, and actually appear to favor going ahead

---

[7] See Part 2, Chap. 3, para. 3, above.

with the liberalization process and opening their airlines to foreign interests in order to strengthen the airline industry.

The present analysis suggests that States collaborate more with each other, on a global basis. International cooperation has always been more efficient than protectionism in improving the national industries. Today, increased cooperation, through cross-border investments, will undoubtedly contribute to the growth and expansion of the airline industry. The judgment of the European Court of Justice, delivered on 5 November 2002, as regards the Open-Skies cases, is a step towards more concentration in the airline industry. Indeed, the condemnation by the Court of the 'nationality clause' will kick off a restructuring of the European airline industry, through mergers and takeovers. It remains that such restructuring will occur only once third countries recognize the Community clause, and consequently, do not threaten the European airlines to get back the traffic rights previously granted. Moreover, the Court recognized an exclusive external competence to the Commission as regards several issues, i.e. pricing and CRS. Although the Commission does not have yet an exclusive competence to negotiate bilateral agreements, Member States cannot negotiate individually anymore without taking into account the Commission's position. Accordingly, it is now highly desirable that the ECJ judgment drives the Council to grant a complete mandate to the Commission to negotiate with third countries on behalf of the European Member States. Such a mandate would allow the EU and the US to attempt to find a compromise on foreign investments in the TCAA framework, despite some unresolved issues. It would certainly solidify a base of common agreements. It is only at this point that it will be possible to discuss the idea of drafting a multilateral convention that deals with subjects such as foreign ownership and cabotage.

In the meantime, the relevant international organizations should coordinate their activities aimed at air transport liberalization. The question of which organizations should be in charge of the process is clearly irrelevant. ICAO, WTO and OECD are complementary, and have their role to play, though not in isolation. Furthermore, these organizations adhere to the same basic principles, such as fairness and transparency, and they pursue the same goals, which in this case is liberalizing international air transport in the public national, regional and global interest. Since their roles are different, it is their responsibility, *vis-à-vis* the international community, to coordinate and join forces in order to turn the abolition of national ownership and control restrictions into reality. This coordination would not, by itself, guarantee the health of the airline industry, but it would finally provide that industry with the same tools most other industries have at their disposal. It would then turn aviation into a normal industry, for better or worse.

# Bibliography

## 1. PRIMARY MATERIALS

### 1.1 Legislation

#### 1.1.1 Canada

Air Canada Public Participation Act, An Act to Provide for the Continuance of Air Canada under the Canada Business Corporations Act and for the Issuance and Sale of Shares thereof to the Public, Assented to August 18, 1988, Chapter A-10.1.

Canadian Transportation Act, An Act to Continue the National Transportation Agency as the Canadian Transportation Agency, to Consolidate and Revise the National Transportation Act, 1987, and the Railway Act and to Amend or Repel Other Acts as a Consequence, Assented to May 29th, 1996, Chapter C-10.4.

*Competition Act*, RS, 1985, c. C-34, s 1; RS, 1985, c. 19 (2nd Supp.), s. 19.

*National Transportation Act*, 1987, R.S.C. 1985, c. 28 (3rd Supp.).

#### 1.1.2 France

*Loi sur les Sociétés Commerciales*, no. 66-537 du 24 juillet 1966, Article 355-1 al.1 (modifiée par la loi no. 85-705 du 12 Juillet 1985).

#### 1.1.3 United States

*Air Commerce Act*, Pub. L. No. 69-254, SS 1-14, ch. 344, § 3(a), 44 Stat. 568, 569, and § 9(a), 44 Stat. 573 (1926).

*Airline Deregulation Act*, Pub. L; No. 95-904, § 102(7), (10), 92 Stat. 1705 (codified as amended at 49 USC § 1301-1552 (1982)).

*Bill to Amend Title 49, United-States Code, to Authorize the Secretary of Transportation to Reduce under Certain Circumstances the Percentage of Voting Interests of Air Carriers which are Required to be Owned and Controlled by Persons who are Citizens of the United-States*, 104th Congress 1st Session, H.R. 951 (15 February 1995).

*Civil Aeronautics Act*, Pub. L. No. 75-706, ch. 601, § 1(13), 52 Stat. at 978 (1938).

*International Antitrust Enforcement Act*, 15 U.S.C. §§ 6201-6212 (1994).

*Maritime Security Act*, Pub. L. No. 104-239, 110 Stat. 3118 (1996).

*Securities Exchange Act*, 17 C.F.R. § 240.12b-2 (1988 & Supp. 1995).

*Sherman Antitrust Act*, 15 U.S.C. §§ 1-2 (1994).

## 1.2 Cases

Air Transport Committee, *Okanagan Helicopters Ltd. Change of Control*, Decision No. 7791 (15 December 1983).

*ECJ, Ahmed Saeed Flugreisen and Silver Line Reiseburo GmbH v. Zentrale zur Bekampfung unlauteren Wettbewenbs e. V. Reference for a Preliminary Ruling: Bundesgerichtshof - Germany . Competition – Air Tariffs*, C- 66/86 [1989] E.C.R. 803, 822.

*ECJ, Commission of the European Communities v. United Kingdom of Great Britain and North Ireland* (C-466/98), *Commission of the European Communities v. Kingdom of Denmark* (C-467/98), *Commission of the European Communities v. Kingdom of Sweden* (C-468/98), *Commission of the European Communities v. Republic of Finland* (C-469/98), *Commission of the European Communities v. Kingdom of Belgium* (C-471/98), *Commission of the European Communities v. Grand Duchy of Luxembourg* (C-472/98), *Commission of the European Communities v. Federal Republic of Austria* (C-475/98), *Commission of the European Communities v. Federal Republic of Germany* (C-476/98), online europa http://curia.eu.int/ (date accessed: 5 November 2002).

National Transport Agency, *Air 2000 Airlines*, decision No. 239-A-1988 (12 August 1988).

National Transport Agency, *Minerve Canada*, Decision No. 618-A-1989 (6 December 1989). online: Canadian Transportation Agency http://www.cta-otc.gc.ca/decisions/1989/A/618-A-1989_e.html (date accessed: 4 October 2001).

National Transport Agency, *Canadian Airlines International Ltd.*, Decision No. 297-A-1993 (May 27, 1993), online: Canadian Transportation Agency http://www.cta-otc.gc.ca/rulings-decisions/decisions/1993/A/297-A-1993_e.html (date accessed: 4 October 2001).

## 1.3 Government Documents

### 1.3.1 Canada

Standing Committee on Transport, *Restructuring Canada's airline industry: fostering competition and protecting the public interest*, report (December 1999), online: Canada's Parliament http://www.parl.gc.ca/InfoComDoc/36/2/TRAN/Studies/Reports/tranrp01/09-rap-e.htm (date accessed: 11 May 2001).

### 1.3.2 United States

Civil Aeronautics Board, *Order in the Matter of Willye Peter Daetwyler, D.B.A. Interamerican Airfreight Co., for Amendment of its Foreign Permit Pursuant to Section 402(f) of the FAA of 1958*, Docket No. 118, 120-21 (1971).

Civil Aeronautics Board, *Order in the Matter of Première Airlines, Fitness Investigation*, CAB Order No. 82-5-11 (5 May 1982).

Department of Transportation, *Application of DHL Aiways, Inc. Pursuant to 49 U.S.C. Section 40109(c) – Exemption – U.S.-Kuwait via Brussels and Bahrain*, Docket No. OST-2000-6937 (14 February 2000), online: DOT http://dms.dot.gov/search/hitlist.asp (date accessed: 4 October 2001).

Department of Transportation, *Application of the Registration of DHL Worldwide Express, Inc., as a Foreign Air Freight Forwarder*, Docket No. OST-2000-8732-1 (10 October 2000), online: DOT http://dms.dot.gov/search/hitlist.asp (date accessed: 4 October 2001).

Department of Transportation, *Application of Federal Express Corporation against DHL Airways, Inc. Regarding Compliance with U.S. Citizenship*, Docket No. OST-2001-8736 (19 January 2001), online: DOT http://dms.dot.gov/search/hitlist.asp (date accessed: 4 October 2001).

Department of Transportation, *Entry and Competition in the U.S. Airline Industry: Issues and Opportunities*, Special Report 255 (30 July 1999), online: DOT http://www.ostpxweb.dot.gov/aviation/domau/dottrbre.pdf (date accessed: 10 May 2001).

Department of Transportation, *Order in the Matter of Page Avjet Corporation*, Order 83-7-5, Docket No. 40,905 (1 July 1983).

Department of Transportation, *Order in the Matter of Intera Arctic Services, Inc.*, DOT Order 87-8-43, Docket No. 44,723 (18 August 1987).

Department of Transportation, *Order in the Matter of the Acquisition of Northwest Airlines by Wings Holdings, Inc.*, DOT Order 89-9-29, Docket No. 46371 (29 September 1989).

Department of Transportation, *Order in the Matter of the Acquisition of Northwest Airlines by Wings Holdings, Inc.*, DOT Order 91-1-41 (14 January 1991).

Department of Transportation, *Order in the Matter of Defining 'Open-Skies'*, 57 Fed. Reg. 19323-01, DOT Order No. 92-4-53 (5 May 1992).

Department of Transportation, *Order in the Matter of Defining 'Open Skies'*, DOT Order 92-8-13, Docket 48130 (5 August 1992).

Department of Transportation, *Order in the Matter of Joint Application of British Airways PLC for an Exemption Pursuant to Section 416(b) of the Federal Aviation Act of 1958; Application of USAir for a Statement of Authorization to Offer Code-Share under 14 CFR Parts 207 and 212; Application of USAir for a Statement of Authorization for a Wet Lease*, DOT Order 93-3-17, Docket Nos. 48,634, 48,640 (15 March 1993).

Department of Transportation, *Order in the Matter of Amending the US Licenses of Carriers of the Participating Countries*, DOT Order 01-1-13 (16 January 2001).

Department of Transportation, *Order Dismissing Third-Party Complaint of Federal Express Corporation in Docket OST-t-2001-8736 and of United Parcel Service Co. (in Docket OST-2001-8824) Without Prejudice Grant the Motions to File Otherwise Unauthorized Documents Filed by Federal Express Corporation and DHL Airways Inc.*, DOT Order 2001-5-11, Docket OST-01-8736-8 (11 May 2001), online: DOT http://152.119.239.10/docimages/pdf58/120921 web.pdf (date accessed: 14 May 2001).

The National Commission to Ensure a Strong Competitive Airline Industry, *Change, Challenge, and Competition: a Report to the President and Congress submitted on 19 August* 1993, Washington, DC: US Government Printing Office (1993).

United States General Accounting Office, *Airline Competition. Impact of Changing Foreign Investment and Control Limits on U.S. Airlines in Report to Congressional Requesters*, GAO Doc. GAO/RCED-93-7 (9 December 1992).


1.3.3 Australia

Australian Commonwealth Department of transport and regional services, *International Air Services*, Policy Statement (June 2000), online: the Australian Commonwealth Department of Transport and Regional Services http://www.dotrs.gov.au/aviation/intairservices.pdf (date accessed: 10 May 2001).

**1.4 International Materials**

1.4.1 Treaties and Other International Agreements

*Additional Protocol N°1 to Amend the Convention for the Unification of Certain Rules relating to International Carriage by Air signed at Warsaw on 12 October 1929*, signed at Montreal, 25 September 1975, ICAO Doc. 9145, 22 I.L.M. 13 (not yet in force).

*Additional Protocol N°2 to Amend the Convention for the Unification of Certain Rules relating to International Carriage by Air signed at Warsaw on 12 October 1929*, signed at Montreal, 25 September 1975, ICAO Doc. 9146, 22 I.L.M. 13 (not yet in force).

*Additional Protocol N°3 to Amend the Convention for the Unification of Certain Rules relating to International Carriage by Air signed at Warsaw on 12 October 1929*, signed at Montreal, 25 September 1975, ICAO Doc. 9147, 22 I.L.M. 13 (not yet in force).

*Additional Protocol N°4, to Amend the Convention for the Unification of Certain Rules relating to International Carriage by Air signed at Warsaw on 12 October 1929*, signed at Montreal, 25 September 1975, ICAO Doc. 9148, 22 I.L.M. 13 (entered into force 14 June 1998).

*Agreement Between the Government of the United States of America and the Government of the United Kingdom Related to Air Services Between their Respective Territories*, 11 February 1946, U.S.-U.K., 60 Stat. 1499.

*Agreement Between the Government of the United States of America and the Government of the United Kingdom Related to Air Services Between their Respective Territories*, 23 July 1977, 28 U.S.T. 5367.

*Agreement Between the United-States of America and the Netherlands Amending the Agreement of April 3, 1957, as Amended and the Protocol of March 31, 1978, as Amended*, 14 October 1992, US-Neth., T.IA.S. 11976.

*Agreement Between the Government of the United-States of America and the Commission of the European Communities Regarding the Application of their Competitive Laws*, O.J. (1995) L. 95/47.

*Air Transport Agreement Between the Federal Republic of Germany and Brunei Darussalam*, German Federal Gazette (BGB1) 1994, II-3670.

*Convention for the Unification of Certain Rules Relating to International Carriage by air*, signed at Warsaw on 12 October 1929, 137 L.N.T.S. 11, 49 Stat. 3000, T.S. 876, ICAO Doc. 601 (entered into force on 13 February 1933).

*Convention for the Unification of Certain Rules for International Carriage by Air*, signed at Montreal, 28 May 1999, ICAO DCW Doc. No. 57 (not yet in force).

*Convention on International Civil Aviation*, 7 December 1944, 15 U.N.T.S. 295, ICAO Doc. 7300/6.

*Convention Supplementary to the Warsaw Convention, for the Unification of Certain Rules Relating to International Carriage by Air Performed by a Person Other than the Contracting Carrier*, signed in Guadalajara, 18 September 1961, 500 U.N.T.S. 31, ICAO Doc. 8181 (entered into force 1 May 1964)

*International Air Services Transit Agreement*, 7 December 1944, 84 U.N.T.S. 389, 394, ICAO Doc. 7500, also reproduced in ICAO Doc. 9587.

*International Air Transport Agreement*, 7 December 1944, 171 U.N.T.S. 387.

*Marrakesh Agreement Establishing the World Trade Organization*, 33 I.L.M. 1144 (entered into force 1 January 1995)).

*Multilateral Agreement on the Liberalization of International Air Transportation*, 15 November 2000 (opened for signature).

*North American Free Trade Agreement Between the Government of Canada, the Government of Mexico and the Government of the United-States*, 17 December 1992, Can. T.S. 1994 No. 2 (1993) 32 I.L.M. 289 (entered into force 1 January 1994).

*Protocol to Amend the Convention for the Unification of Certain Rules relating to International Carriage by Air signed at Warsaw on 12 October 1929*, done at The Hague, 28 September 1955, 478 U.N.T.S. 371, ICAO Doc. 7632 (entered into force on 1 August 1963).

*Protocol to Amend the Convention for the Unification of Certain Rules relating to International Carriage by Air signed at Warsaw on 12 October 1929, as Amended by Protocol done at The Hagueon 28 September 1955*, signed at Guatemala City, 8 March 1971, ICAO Doc. 8932 (not yet in force).

*Treaty of Amsterdam, Amending the Treaty on European Union, the Treaties Establishing the European Communities and Certain Related Acts*, 2 October 1997, O.J. (C340)1(1997).

1.4.2 International Civil Aviation Organization Documents

ICAO, *Annual Report of the Council - 2000*, ICAO Doc. 9770 (2001).

ICAO, *Broadening of Ownership and Control Criteria*, ICAO Doc. ATRP/WP/5 (14 November 1996).

ICAO, *Questionnaire on State's Policies and Practices Concerning Air Carrier Ownership and Control*, Attachment to State letter SC 5/2-01/50, online: ICAO http://www.icao.int/cgi/goto_atb.pl?icao/en/atb/ecp/S150-survey.htm;ecp (date accessed: 2 January 2001).

ICAO, *Manual on the Regulation of International Air Transport*, ICAO Doc 9626 (1$^{st}$ ed.) (1996).

ICAO, *Policy and Guidance Material on the Economic Regulation of International Air Transport*, ICAO Doc. 9587 (1999).

ICAO, *Report of the Air Transport Regulation Panel Working Group on Air Carrier Ownership and Control*, ICAO doc. (September 2002), online: ICAO http://www.icao.int/atrp (date accessed: 1 October 2002).

ICAO, *The World of Civil Aviation, 2000 – 2003* (Provisional publication of the Circular 287), ICAO Doc. AT/122 (9 October 2001).

ICAO, Working Paper (*World-Wide Air Transport Conference on International Air Transport Regulation: Present & Future*) No. AT Conf/4 – WP 7 (18 April 1994).

ICAO, Working Paper (*World-Wide Air Transport Conference on International Air Transport Regulation: Present & Future*) No. AT Conf/4 – WP 10 (19 April 1994).

ICAO, Working Paper (*World-Wide Air Transport Conference on International Air Transport Regulation: Present & Future*) No. AT Conf/4 – WP 8 (20 April 1994).

ICAO, Working Paper (*World-Wide Air Transport Conference on International Air Transport Regulation: Present & Future*) No. AT Conf/4 – WP 9 (21 April 1994).

ICAO, Working Paper (*World-Wide Air Transport Conference on International Air Transport Regulation: Present & Future*) No. AT Conf/4 – WP 18 (20 July 1994).

ICAO, Working Paper (*World-Wide Air Transport Conference on International Air Transport Regulation: Present & Future*), No. AT Conf/4 – WP 5 (8 August 1994).

ICAO, Working Paper (*World-Wide Air Transport Conference on International Air Transport Regulation: Present & Future*) No. AT Conf/4 – WP 30 (12 August 1994).

ICAO, Working Paper (*World-Wide Air Transport Conference on International Air Transport Regulation: Present & Future*) No. AT Conf/4 – WP 47 (23 August 1994).

ICAO, Working Paper (*World-Wide Air Transport Conference on International Air Transport Regulation: Present & Future*) No. AT Conf/4 – WP 52 (8 September 1994).

ICAO, Working Paper (*World-Wide Air Transport Conference on International Air Transport Regulation: Present & Future*) No. AT Conf/4 – WP 68 (20 October 1994).

ICAO, Working Paper (*Report by the Council on Trade in Services*) No. A31-WP/23, EC/3 (7 April 1995).

ICAO, Working Paper (*Report of the Council on Trade in Services*) No. A33-WP/7 (5 June 2001).

ICAO, Working Paper (*Substantial Ownership and Effective Control of Designated Airlines*) No. A33-WP/96 (17 August 2001).

ICAO, Working Paper (*Substantial Ownership and Effective Control over Designated Airlines*) No. A33-WP/181 (25 September 2001).

ICAO, Working Paper (*The Orderly Evolution of Air Transport Services: Secure and Safe Economic Regulation in an Area of Globalisation*) No. A33-WP/227 (28 September 2001).

ICAO, Working Paper (*Economic Commission – Draft Text for the Report on Agenda Item 26*) No. A33-WP/262 (2 October 2001).

ICAO, Working Paper (Result of the Survey of States' policies and practices concerning air carrier ownership and control) No. AT-WP/1933 (2 April 2002).

ICAO, Working Paper (*Liberalizing Air Carrier Ownership and Control*) No. ATConf/5-WP/7 (21 October 2002).

### 1.4.3 European Union Documents

*1.4.3.1 Regulations, Directives, Decisions*

EU, *Commission Decision No. 95/404/EC on a Procedure Relating to the Application of the Council Regulation 2407/92 (Swissair/Sabena)*, [1995] O.J. L. 239/19.

EU, *Commission Proposal No. 500PC0595 for a Regulation of the European Parliament and of the Council on Establishing Common Rules in the Field of Civil Aviation and Creating a European Aviation Safety Agency*, (2 July 2001), online europa
http://www.europa;eu;int/eur-lex/en/com/dat/2000/en_500PC0595.html (date accessed: 15 January 2002).

EU, *Council Decision 87/602/EEC on the Sharing of Passenger Capacity between Air Carriers on Scheduled Air Services between Member States and on Access for Air Carriers to Scheduled Air Service Routes between Member States*, [1987] O.J. L. 374/19.

EU, *Council Directive 87/601/EEC on Fares for Scheduled Air Services between Member States*, [1987] O.J. L. 374/12.

EU, *Council Regulation 3975/87 Laying Down the Procedure for the Application of the Rules on Competition to Undertakings in the Air Transport Sector*, [1987] O.J. L. 374/1.

EU, *Council Regulation 3976/87 on the Application of Article 85(3) of the Treaty to Certain Categories of Agreements and Concerted Practices in the Air Transport Sector*, [1987] O.J. L. 374/9.

EU, *Council Regulation 4064/89 on the Control of Concentrations Between Undertakings*, [1989] O.J. L. 395/1.

EU, Council Regulation 2342/90 on Fares for Scheduled Air Services, [1990] O.J. L. 217/1.

EU, *Council Regulation 2343/90 on Access for Air Carriers to Scheduled Intracommunity Air Service Routes and on the Sharing of Passenger Capacity between Air Carriers on Scheduled Air Services between Member States*, [1990] O.J. L. 217/8.

EU, *Council Regulation 2344/90 Amending Regulation 3676/87 on the Application of Article 85(3) of the Treaty to Certain Categories of Agreements and Concerted Practices in the Air Transport Sector*, [1990] O.J. L. 217/15.

EU, *Council Regulation 2407/92 on Licensing of Air Carriers*, [1992] O.J. L. 240/1.

EU, *Council Regulation 2408/92 on Access for Community Air Carriers to Intracommunity Air Routes*, [1992] O.J. L. 240/8 (corrected in [1992] O.J. L. 15/33).

EU, *Council Regulation 2409/92 on Fares and Rates for Air Services*, [1992] O.J. L. 240/15.

EU, *Council Regulation 2410/92 Amending Regulation 3975/87 Laying Down the Procedure for the Application of the Rules to Competition to Undertakings in the Air Transport Sector*, [1992] O.J. L. 240/18.

EU, *Council Regulation 2411/92 Amending Regulation 3976/87 on the Application of Article 85(3) of the Treaty to Certain Categories of Agreements and Concerted Practices in the Air Transport Sector*, [1992] O.J. L. 240/19.

EU, *Council Regulation 95/93 on Common Rules for the Allocation of Slots at Community Airports*, [1993] O.J. L.14/1.

EU, *Council Regulation 3089/93 Amending Regulation 2299/89 on a Code of Conduct for Computerized Reservation Systems*, [1993] O.J. L.278/1.

EU, *Decision of the EEA Joint Committee 7/94 Amending the Protocol 47 and Certain Annexes to the EEA Agreement*, [1994] O.J. L. 160/1.

### 1.4.3.2 Other European Documents

EU, *European Commission, Directorate General for Energy and Transport document* 'Air Security – Short presentation of the proposal for a regulation establishing common rules for civil aviation security' (October 2001), online: europa http://www.europa.eu.int/common/transport/library/press-kit-surete-en.pdf (date accessed: 6 December 2001).

EU, *Commission Communication on the Consequences of the Court Judgment of 5 November 2002 for European Air Transport Policy*, COM(2002) 649 final, 19 November 2002.

EU, 'Towards new rules on aviation security following the attacks' Doc. IP/01/1397 (10 October 2001), online: europa http://www.europa.eu.int/rapid/start/cgi/guesten.ksh?p_action.gettxt=gt&doc=IP/0 1/1397 (date accessed: 6 December 2001).

EU, *Council meeting N° 2472 on Transport, Telecommunications and Energy*, 15121/02 (Press 380) 5-6 December 2002.

1.4.4 Organization for Economic Cooperation and Development Documents

OECD, Directorate for Science, Technology, and Industry – Division of Transport, *Regulatory Reform in International Air Cargo Transportation*, Doc. No. DSTI/DOT(99)1 (June 1999).

OECD, Directorate for Science, Technology, and Industry – Division of Transport, *OECD Principles for the Liberalization of Air Cargo*, Doc. No. DSTI/DOT(2000)1 (June 2000).

OECD, Directorate for Science, Technology, and Industry – Division of Transport, *Draft Annotated Agenda – OECD Workshop on Principles for the Liberalization of Air Cargo Transportation*, Doc. No. DSTI/DOT/A(2000)1 (August 2000).

OECD, Directorate for Science, Technology, and Industry – Division of Transport, *OECD Principles for the Liberalization of Air Cargo Transportation – Comments on DSTI/DOT(2000)1*, Doc. No. DSTI/DOT/RD(2000)1 (September 2000).

OECD, Directorate for Science, Technology, and Industry – Division of Transport, *OECD Principles for the Liberalization of Air Cargo Transportation – Comments on Articles contained in DSTI/DOT(2000)1*, Doc. No. DSTI/DOT/RD(2000)2 (September 2000).

OECD, Directorate for Science, Technology, and Industry – Division of Transport, *OECD Workshop on Principles for the Liberalization of Air Cargo Transportation – Paris, 4-5 October 2000 – Summary Record*, Doc. No. DSTI/DOT/M(2000)1 (November 2000).

OECD, Directorate for Science, Technology, and Industry – Division of Transport, *Liberalization of Air Cargo Transport*, Doc. No. DSTI/DOT(2002)1/REV1 (May 2002).

1.4.5 Documents from Other International Organizations

AEA, *Towards a Transatlantic Common Aviation Area*, AEA Policy Statement (September 1999).

ECAC, *Report on Task Force on Ownership and Control Issues, First Meeting*, ECAC Doc. OWNCO/1 (24 December 1998).

ECAC, *Report on Task Force on Ownership and Control Issues, Second Meeting*, ECAC Doc. OWNCO/2 (26 February 1999).

IATA, Government and Industry Affairs Department, *Report of the Ownership & Control Think Tank World Aviation Regulatory Monitor*, IATA doc. prepared by H.P. van Fenema (7 September 2000)

WTO, *Note on Developments in the Air Transport sector Since the Conclusion of the Uruguay Round*, Part Four. WTO Doc. S/C/W/163/Add.3 (2001).

WTO, *Note on Developments in the Air Transport sector Since the Conclusion of the Uruguay Round*, Part Five. WTO Doc. S/C/W/163/Add.4 (2001).

## 2. SECONDARY MATERIALS

### 2.1 Books

Abeyratne, R.I.R., *Emergent Commercial Trends and Aviation Safety* (Aldershot: Ashgate, 1999).

Abeyratne, R.I.R., *Aviation Trends in the New Millennium* (Aldershot: Ashgate, 2001).

Flory, T., *L'Organisation Mondiale du Commerce – Droit institutionnel et substantiel* (Bruxelles: Etablissements Emile Bruylant, 1999).

Hill, C.W.L., *International Business: Competing in the Global Marketplace*, 2nd ed. (Chicago: Richard D. Irwin, 1997).

Hill, C.W.L., *International Business: competing in the global marketplace*, 3rd ed. (Boston: Irwin/McGraw-Hill, 2001).

Hufbauer, G.C., Findlay, C., *Flying High – Liberalizing Civil Aviation in the Asia Pacific* (Washington DC: Institute for International Economics, 1996).

Jackson, J.H., *The World Trading System – Law and Policy of International Economic Relations*, 2nd ed. (Cambridge: the MIT Press, 1997).

Johnston, D., Johnston, D., Handa, S., *Getting Canada Online – Understanding the Information Highway* (Toronto: Stoddart Publishing Co. Limited, 1995).

Lu, A. C., *International Airline Alliances : EC Competition Law/ US Antitrust Law and International Air Transport* (The Hague: Kluwer Law International, 2003).

Trebilcock, M.J., Howse, R., *Regulation of International Trade*, 2nd ed. (London: Routledge, 1999).

### 2.2 Articles in Journals

Abeyratne, R.I.R., 'Would Competition in Commercial Aviation ever Fit into the WTO?' (1996) 61 J. Air L. & Com. 793.

Abeyratne, R.I.R., 'Emergent Trends in Aviation Competition Laws in Europe and in North America' (2000) 23 World Competition. R. 141.

Alexandrakis, C.G., 'Foreign Investment in U.S. Airlines: Restrictive Law is Ripe for Change' (1994) 4 U. Bus. Miami L.J. 71.

American Bar Association, 'Cross-Border Investment in International Airlines: Presenting the Issues' (2000) Air & Space Law. 20.

Arlington, D.T., 'Liberalization of Restrictions on Foreign Ownership in U.S. Air Carriers: the United States must take the First Step in Aviation Globalization' (1993) 59 J. Air L. & Com. 133.

AuBuchon, M.J., 'Testing the Limits of Federal Tolerance: Strategic Alliances in the Airline Industry' (1999) 26 Transp. L.J. 219.

Balfour, J., 'Airline Mergers and Marketing Alliances – Legal Constraints' (1995) 20 Air & Space L. 112.

Balfour, J., 'A Question of Competence: the Battle for Control of European Aviation Agreements with the US' (2001) 16-SUM Air & Space Law. 7.

Basedow, J., 'Airline Deregulation in the European Community – its Background, its Flaws, its consequences for E.C.-U.S. Relations' (1994) 13 J.L. & Com. 247.

Binggeli, U., Pompeo, L., 'Hyped hopes for Europe's law-cost airlines' (2002) The McKinsey Quarterly 4, online The McKinsey Quarterly http://www.mckinseyquarterly.com/article_page.asp?ar=1231&L2=23&L3=79 &srid= (date accessed: 30 October 2002).

Bliss, F.A., 'Rethinking Restrictions on Cabotage: Moving to Free Trade in Passenger Aviation' (1994) 17 Suffolk Transnat'l L. Rev. 382.

Bohmann, K., 'The Ownership and Control Requirement in U.S. and European Union Air Law and U.S. Maritime Law-Policy; Consideration; Comparison' (2001) 66 J.Air L. & Com. 689.

Brenner, M. 'Airline Deregulation – A Case Study in Public Policy Failure' (1988) 16 Transp. L. J. 179.

Brown, J.D., 'Foreign Investment in U.S. Airlines: What Limits should be Placed on Foreign Ownership of U.S. Carriers?' (1990) 41 Syracuse L.R. 1269.

'Complaints About U.S. Airlines Up Almost Seven-Fold In August', (1987) J. Rec. (Okla. City).

Costa, P., Harned, D., Lundquist, J., 'Rethinking the aviation industry' (2002) The McKinsey Quarterly 2, online The McKinsey Quarterly http://www.mckinseyquarterly.com/article_page.asp?ar=1190&L2=23&L3=79 (date accessed: 13 May 2002).

Dempsey, S., 'Airlines in Turbulence: Strategies for Survival' (1995) 23:15 Transp. L. J. 15.

Dempsey, S., 'Competition in the Air: European Union Regulation of Commercial Aviation' (2001) 66 J. Air L.& Com. 979.

Doganis, R., 'Relaxing Airline Ownership and Investment Rules' (1996) 21 Air & Space L. 267.

Edwards, A., 'Foreign Investment in the U.S. Airline Industry: Friend or Foe?' (1995) 9 Emory Int'l L.R. 595.

Elliott, G.P., 'Antitrust at 35,000 Feet: the Extraterritorial Application of United States and European Community Competition Law in the Air Transport Sector' (1997-1998) 31 Geo. Wash. J. Int'l L. & Econ. 185.

Gawlicki, S., 'Virtual Mergers: With Traditional Mergers Difficult to Pull off, Airlines Finding Creative Ways to Consolidate' (2000) Inv. Dealers' Dig. (WL 4666779).

Gertler, J.S., 'Nationality of Airlines: Is It a Janus with Two (or More) Faces?' (1994) 19:1 A.A.S.L. 211.

Goo, G.L.H., 'Deregulation and Liberalization of Air Transport in the Pacific Rim: Are They Ready for America's "Open Skies"?' (1996) 18 U. Haw. L. Rev. 541.

Gorton, L., 'Air Transport and EC Competition Law' (1998) 21 Fordham Int'l L.J. 602.

Grant, Th.D., 'Foreign Takeovers of United States Airlines: Free Trade Process, Problems and Progress' (1994) 31 Harv. J. on Legis. 63.

Haanappel, P.P.C., 'Airline Challenges: Mergers, take-overs, alliances and franchises' (1995) 21 A.A.S.L. 179.

Haanappel, P.P.C., 'Airline Ownership and Control, and Some Related Matters' (2001) 26-2 Air & Space L. 90

Janda, R., 'Passing the Torch: Why ICAO Should Leave Economic Regulation of International Air Transport to the WTO' (1995) 21 Air & Space L. 409

Karber, P., 'Re-constructing Global Aviation in an Era of the  Civil Aircraft as a Weapon of Destruction' (2002) 25 Harv. J. L. & Pub. Pol'y 781.

Kass, H.E., 'Cabotage and Control: Bringing U.S. Aviation Policy into the Jet Age' (1994) 26 Case W. Res. J. Int'l L. 143.

Lehner, R.D., 'Protectionism, Prestige, and National Security: The Alliance Against Multilateral Trade in International Air Transport' (1995) 45 Duke L.J. 436.

Little, V., 'Control of International Air Transport' (1949) III International Organization 29.

Lipman, G., 'Multilateral Liberalization – The Travel and Tourism Dimension' (1994) 19 Air & Space L. 152.

Luz, K., 'The Boeing-McDonnel Douglas Merger: Competition Law, Parochialism, and the Need for a Globalized Antitrust System' (1999) 32 Geo. Wash. J. Int'l L.& Econ. 155.

Milde, M., 'The Chicago Convention – Are Major Amendments Necessary or Desirable 50 Years Later' (1994) 19:1 A.A.S.L. 401.

Milde, M., 'Enforcement of Aviation Safety Standards' (1996) 45 Abhandlungen 3.

Miller, L., 'Airline Merger Offers Fliers No Pie in Sky' (1996) Wall St. J. Eur. 8.

Mosin, S., 'Riding the Merger Wave: Strategic Alliances in the Airline Industry' (2000) 27 Transp. L.J. 271.

Mosteller, J., 'The Current and Future Climate of Airline Consolidation: The Possible Impact of an Alliance of Two Large Airlines and an Examination of The Proposed American Airlines-British Airways Alliance' (1999) 64 J.Air L. & Com. 575.

Naveau, J., 'Les Alliances entre Compagnies Aériennes. Aspects Juridiques et Conséquences sur l'Organisation du Secteur' (1999) 49 ITA Etudes & Doc. 9.

O'Toole, T., '"The Long Arm of the Law" – European Merger Regulation and its Application to the Merger of Boeing & McDonnell Douglas' (1998) 11 Transnat'l L. 203.

'Ownership Trend Creates Need for New Links Between States and Airlines' (June 1992) 47 ICAO J. 14.

Perkins, E., 'Mergers will Squeeze Consumers' (1997) Orange County (Cal.) Reg. D04.

Polley, R., 'Defense Strategies of National Carriers' (2000) 23 Fordham Int'l L.J. 170.

'Progressive Liberalization Actively Supported by ICAO, Council President Tells IATA Annual General Meeting' (June 2001) 56 ICAO J. 30.

Rodriguez Serrano, V., 'Trade in Air Transport Services: Liberalizing Hard Rights' (1999) 24 A.A.S.L. 199.

Schless, A.L., 'Open Skies: Loosening the Protectionist Grip on International Civil Aviation' (1994) 8 Emory Int'l L. Rev. 435.

Schulte-Strathaus, U., 'Common Aviation Areas: the Next Step Toward International Air Liberalization' (2001) 16-SUM Air & Space Law. 4.

Simon, M.S., 'Aviation Alliances: Implications for the Quantas – BA Alliance in the Asia Pacific Region' (1997) 62 J. Air L. & Com. 841

Sorensen, F., van Weert, W., Lu, A.C., 'ECJ Ruling on Open Skies Agreements v. Future International Air Transport Relations' (2003) Air & Space L. (Forthcoming).

Stewart, J.T., 'US Citizenship Requirements of the Federal Aviation Act – A Misty Moor of Legalisms or the Rampart of Protectionism' (1990) 55 J. Air L. & Com. 685

Swinnen, B.M.J., 'An Opportunity for Transatlantic Civil Aviation: from Open Skies to Open Markets?' (1997) 63 J. Air L. & Com. 249.

Tiwari, S., Chik, W.B., 'Legal Implications of Airline Cooperation: Some Legal Issues and Consequences Arising from the Rise of Airline Strategic Alliances and Integration in the International Dimension' (2001) J. of Aviation Management of Singapore Aviation Academy 9.

van Fenema, H.P., 'Substantial Ownership and Effective Control as Airpolitical Criteria' (1992) Air & Space Law: De Lege Ferenda (Liber Amicorum Henri Wassenbergh), Masson-Zwaan and Mendes de Leon eds., pp. 27-41, the Netherlands.

van Fenema, H.P., 'Ownership Restrictions: Consequences and Steps to be Taken' (1998) 23 Air & Space L. 63.

van Traa-Engelman, H.L., 'Reports of Conferences: The European Air Transport Conference – Airline Globalization' (1998) 23:1 Air & Space L. 31.

Warner, S.M., 'Liberalize Open Skies : Foreign Investment and Cabotage Restrictions Keep Non Citizens in Second Class' (1993) 43 Am. U. L. Rev. (Washington, DC) 277.

Wassenbergh, H.A., 'Future Regulation to Allow Multi-National Arrangements Between Air Carriers (Cross-Border Alliances), Putting an End to Air Carrier Nationalism' (1995) 20 Air & Space L. 164.

Wassenbergh, H.A., 'The Sixth Freedom Revisited' (1996) 21 Air & Space L. 285.

Whalen, T.J., 'The New Warsaw Convention: the Montreal Convention' (2000) 25:1 Air & Space L. 12.

Young, J.W., 'Globalism Versus Extraterritoriality. Consensus Versus Unilateralism : Is There a Common Ground? A US Perspective' (1999) 24 Air & Space L. 209.

## 1.2 Unpublished Manuscripts

Haghighi, S.S., A Proposal for an Agreement on Investment in the Framework of the World Trade Organization (LL.M. thesis, Institute of Comparative Law, McGill University 1999).

Petras, C.M., 'Foreign Ownership of US Airlines and the Civil Reserve Air Fleet Program: Cause for Concern?' (15 March 2001).

Wassenbergh, H., 'Towards Global Economic Regulation of International Air Transportation through Inter-Regional Bilateralism' The Hague (August 2001).

## 1.3 Addresses and Papers Delivered at Conferences

Al Dawoodi, A.J., 'The impact of the GATS on Air Transport from the General Perspective of Developing Countries' (Dialogue on Trade in Aviation-Related Services, ICAO, Montreal, 12 June 2001) [unpublished].

Janda, R., 'ICAO as a Trustee for Global Public Goods in Air Transport' (Dialogue on Trade in Aviation-Related Services, ICAO, Montreal, 12 June 2001) [unpublished].

Loughlin, R., 'The Current GATS Round in a Historical Perspective' (Dialogue on Trade in Aviation-Related Services, ICAO, Montreal, 12 June 2001) [unpublished].

Palacio (de), L., 'Beyond Open Skies' (The European Commission Beyond Open Skies Conference, Chicago, 6 December 1999).

Palacio (de), L., 'Globalization – The Way Forward' (IATA World Transport Summit, Madrid, 27-29 May, 2001).

Shane J., Association Deputy Secretary, DOT, 'Open Skies Agreements and the European Court of Justice' (American Bar Association, Forum on Air and Space Law, Hollywood Florida, 8 November 2002).

Sørensen, F., 'Market Access Liberalization of Air Transport' (Dialogue on Trade in Aviation-Related Services, ICAO, Montreal, 12 June 2001) [unpublished].

van Fenema, H.P., 'Airline Ownership and Control: Long and Short Term Approaches to a Trade Barrier' (Annual Conference of the European Air Law Association, Zurich, 9 November 2001 / to be published in 2003 by Sakkoulas and Kluwer).

## 1.4 Articles in Magazines

'ANZ Asks Government To Lift Foreign Ownership Limits' *Aviation Daily* 345:10 (16 July 2001) 5.

'APEC Multilateral Moves U.S. Toward Globalizing Pacts' *Aviation Daily* 344:22 (1 May 2001) 3.

'Australian Government to Ease Foreign Ownership Restrictions' *Aviation Daily* (19 August 1999) 3

Baker, C., 'French Push for TCAA' *Airline Bus.* (December 2000) 18.

Baker, C., 'US and UK Remain Apart on Open Skies' *Airline Bus.* (December 2000) 19.

Baker, C., 'History Lessons' *Airline Bus.* (December 2000) 74.

Baker, C., 'Behind The Handshake' *Airline Bus.* (February 2001) 66.

Cameron, D., 'Out of the Wilderness' *Airline Bus.* (June 1999) 50.

Chuter, A., 'Growing pains: Differing Approaches Taken by US and European Regulators Over the Proposed GE/Honeywell Merger Highlight a Need for Common Guidelines' *Flight Int'l* (19 June 2001) 5.

Conway, P., 'Could Cargo Lead Liberalisation' *Airline Bus.* (December 2000) 29.

Feldman, J.F., 'No Guts, no Glory' *Air Transport World* (January 1992) 64.

Feldman, J.M., 'It's Still a Bilateral World' *Air Transport World* (August 1997) 35.

Fennes, R., of DG VII, is quoted by Karen Walker, 'The Great Global Debate' *Airline Bus.* (September 1999) 96.

Field, D., 'Regulatory Hurdles Remain for United's Merger Plans' *Airline Bus.* (April 2001) 13.

Fiorino, F., 'More Open Skies' *Aviation Wk & Space Tech.* (12 June 2000) 19.

Gill, T., 'Opening Arab Skies' *Airline Bus.* (June 1999) 47.

Gill, T., 'A Firmer Base' *Airline Bus.* (June 2000) 49.

Hughes, D., 'ICAO Delegates Shun US Free-Market Stance' *Aviation Wk & Space Tech.* (2 January 1995) 37.

Ionides, N., 'Expanded Horizons' *Airline Bus.* (November 1999) 34.

Ionides, N., 'Spoiling for Choice' *Airline Bus.* (October 2000) 84.

Ionides, N., 'China Merger Takes Shape' *Airline Bus.* (February 2001) 26.

Ionides, N., 'China to Loosen Central Ownership' *Airline Bus.* (April 2001) 27.

Ionides, N., 'Five Sign Up to Asia-Pacific Multilateral Agreement' *Airline Bus.* (June 2001) 34.

'Jeanniot Sees Regional Blocs As Cure For Over-Fragmentation' *Aviation Daily* 344:43 (31 May 2001) 2.

Jones, L., 'When the Going Gets Tough...' *Airline Bus.* (May 1998) 26.

Katz, R., 'New Directions?' *Airline Bus.* (June 1992) 36.

Knibb, D., 'No Flag in its Future' *Airline Bus.* (May 1999) 72.

Knibb, D., 'Australian Ownership Rules Criticized' *Airline Bus.* (August 1999) 26.

Knibb, D., 'Aerolineas Rescue Relies on Spain' *Airline Bus.* (August 2000) 18.

Knibb, D., 'Thai Moves Towards Privatisation' *Airline Bus.* (December 2000) 24.

Knibb, D., 'Bilateral Accord Sparks Ownership Debate...as APEC Moves Towards Multilateral Open Skies' *Airline Bus.* (January 2001) 24.

Lancesseur, B., 'Un Cadre Réglementaire rigide – La mise à plat s'impose' *Aéroports Magazine* (Mai 2001) 18.

Max KJ, 'South African Liberalisation Makes Progress' *Flight Int'l* (24 October 2000) 4.

McKinsey, 'Making Mergers Work' *Airline Bus.* (June 2001) 110.

Montlake, S., 'Sair Takes Portugalia Stake' *Airline Bus.* (August 1999) 20.

Morrocco, J.D., 'Open Skies Impasse Shifts Alliance Plans' *Aviation Wk & Space Tech.* (9 November 1998) 45.

'Multilateral Pact Carriers Must Report Ownership Changes' *Aviation Daily* 342:43 (1 December 2000) 4.

'New Zealand To Allow Increased Foreign Ownership' *Aviation Daily* 342:36 (20 November 2000) 5.

Pilling, M., 'Only a Call Away' *Airline Bus.* (March 2001) 38.

Shifrin, C., 'FAA plans safety change' *Airline Bus.* (June 1999) 11.

Shifrin, C., 'Towards Unsettled Skies' *Airline Bus.* (June 1999) 87.

Sparaco, P., 'European Deregulation Still Lacks Substance' *Aviation Wk & Space Tech.* (9 November 1998) 53

Tarry, C., 'Playing for Profit' *Airline Bus.* (June 1999) 90.

'The Global Alliance Grouping' *Airline Bus.* (May 2000) 59.

'The New Zealand Government Relaxes Foreign Ownership Limits in Air New Zealand' *Air Transport World* (October 2001) 12.

Thornton, C., 'The New: Europe's New Transport Commissioner has Set out her Agenda on Air Transport and Appears Determined to See it Through' *Airline Bus.* (March 2000) 32.

Thornton, C., Lyle, C., 'Freedom's Paths' *Airline Bus.* (March 2000) 74.

'UK Says Airline Merger and Acquisition Rules should be Relaxed' (3 October 2001) Airline Indus. Info.

'US Questions Maersk's Power' *Fair Play* (19 July 2001) 22.

Walker, K., 'US DoD gives Red Light to Ownership Changes' *Airline Bus.* (June 1999) 11.

Walker, K., 'Sans Frontiers?' *Airline Bus.* (February 2000) 34.

Zuckert Scoutt and Rasenberger, 'European Court says "Bye-Bye Bermuda"' *Aviation Advisor*, Special Edition (6 November 2002).

## 1.5 Articles in Newspapers

'Air Jamaica hurt by US tragedy' *The St. Vincent Herald* (4 November 2001), online: Caribbean Alpa http://www.caribbeanalpa.com/news/index.shtml (date accessed: 5 November 2001).

'Air Liberté espère revenir à l'équilibre en 2003' *Le Figaro* (date accessed: 2 November 2001), online : Le Figaro http://www.lefigaro.fr/cgi.bin/gx.cgi/AppLogic+FTContentServer?pagename= FutureTense/Apps/Xcelerate (date accessed: 5 November 2001).

Alexander, K., 'United seeks bankruptcy protection' *Washington Post* (10 December 2002), online: Washington Post http://www.washingtonpost.com/wp-dyn/articles/A32793-2002Dec9.html (date accessed: 11 December 2002).

Bartholomeusz, S., 'Ansett's survival goes to the heart of deregulation policy' *The Age* (7 September 2001), online: The Age http://www.theage.com.au/business/2001/09/07/FFX2JVQF9RC.html (date accessed: 7 September 2001).

Bocev, P., 'l'Europe recale le mariage GE-Honeywell' *Le Figaro* (4 July 2001) 1 and 5.

Brethour, P., 'Telecom ownership review coming: AT&T' *The Globe and Mail* (30 October 2001) B1 & B6.

Coleman, Z., 'Government rescues Air NZ' *The Globe and Mail* (5 October 2001) B7.

Crols, B., 'Flag-carrier Sabena in death spiral' *The Globe and Mail* (6 November 2001) B13.

Damsell, K., 'Ownership rules key: Astral' *The Globe and Mail* (14 December 2001) B5.

Dugua, P.Y., 'Le ciel américain broie du noir' *Le Figaro* (23 June 2001) III.

Evans, G., 'Air NZ shares bounce as talks continue' *The Age* (26 September 2001), online: The Age
http://www.theage.com.au/news/national/2001/09/26/FFX44YQ02SC.html
(date accessed: 26 September 2001).

Gibbens, R., 'World has too many airlines: IATA boss' *National Post* (4 December 2001), online: national Post
http://www.nationalpost.com/financialpost/worldbusiness/story.htm
(date accessed: 5 December 2001).

Glackin, M.,'Airline Consolidation is the only route to survival' *The Scotsman* (24 December 2001), online: The Scotsman
http://209.185.240.250/cgibin/linkrd?Lang=EN&lah=7986709cb8b8d0bfa236
0ae6558c6ed8&lat=1010085559&hmaction=http%3a%2f%2fwww%2eairline
rs%2enet%2fnews%2fredirect%2emain%3fid%3d30291 (date accessed: 24 December 2001).

Guerrera, F., 'GE fires salvo at European Commission' *Financial Post* (5 November 2001) FP11.

Kouamouo, T., 'Air Afrique passe sous la tutelle d'Air France' *Le Monde* (17 Août 2001) A12.

'L'Etat est contraint de jouer au 'pompier social'' *Le Monde* (18 October 2001), online: Le Monde http://www.lemonde.fr/rech_art/0,5987,235581,00.html (date accessed: 5 November 2001).

Mallet, V., 'Companies & finance international: French airlines in crisis' *Financial Times* (9 April 2001), online: Financial Times
http://specials.ft.com/In/ftsurveys/industry/sc22356.htm (date accessed: 16 May 2001).

McArthur, K., 'Ottawa may ease airline ownership rules' *The Globe and Mail* (2 October 2001) A10.

McArthur, K., Chase, S., 'Schwartz Spurns Air Canada as Ottawa Mulls Ownership Cap' *The Globe and Mail* (5 October 2001) B1 and B4.

McArthur, K., 'Air Canada courting investors' *The Globe and Mail* (3 November 2001) B1.

McCartney, S., 'Widening losses at airlines make shake up unavoidable' *The Globe and Mail* (6 November 2001) B15.

Michaels, D., Thurow, R., 'Swiss banks draw ire' *The Globe and Mail* (5 October 2001) B7.

Rose, L.R., & Coleman, B., 'British Airways Buys Stake in USAir, Drawing Protests From Other Carriers' *The Wall St. J.* (22 January 1993) A3.

Thomas, G., 'Air NZ plummet hits SIA and BIL' *The Age* (25 September 2001), online: The Age http://www.theage.com.au/news/national/2001/09/25/FFXEPDK02SC.html (date accessed: 26 September 2001).

'1944 and all that' *Sunday Times – London* (7 October 2001), available on WL 27457432.

**1.6 Press Releases**

Air France, Press Release, 'Lancement du processus de liberalization d'Air France' (29 July 2002), online Air France http://bv.airfrance.fr/cgi-bin/FR/frameset.jsp (date accessed: 20 September 2002).

Council of Canadians, Immediate Release 'Transport Committee Recommendations Threaten Safe, Affordable and Accessible Canadian Air Service' (December 1999), online: the Council of Canadians www.canadians.org/media/media-991208.html (date accessed: 11 May 2001).

Department of Justice, Immediate Release, 'Justice Department Urges DOT to Impose Conditions on American Airlines/British Airways Alliance' (17 December 2001), online: DOJ http://www.usdoj.gov/atr/public/press_releases/2001/9705.htm (date accessed: 18 December 2001).

Department of Transportation, Immediate Release 45-01, 'DOT Rules on Petitions Against DHL' (11 May 2001), online: DOT http://www.dot.gov/affairs/dot45-01.htm (date accessed: 1 October 2001).

Department of Transportation, Immediate Release 68-01, 'US Secretary of Transportation Says Bush Administration to Press for Global Aviation Liberalization' (30 June 2001).

Department of Transportation, Immediate Release 111-01, 'United-States, France Reach Open-Skies Aviation Agreement' (19 October 2001).

DHL WE, Immediate Release , 'DHL Worldwide Express Welcomes DOT Ruling' (11 May 2001), online: DHL WE http://www.dhlusa.com/press_display/1,3574,79,00.html (date accessed: 1 October 2001).

EU, Press Release 16/1998, 'European Commission takes legal action against EU Member States 'Open-Skies' agreements with the United States' (11 March 1998).

ICAO, Immediate Release (PIO 18/2001), 'Safety and Security at the heart of key ICAO achievements for 2001' (27 December 2001), online: ICAO http://www.icao.int/icao/en/nr/pio200118.htm (date accessed: 1 June 2002).

ICAO, Immediate Release PIO 12/2002, 'Council of ICAO Establishes Global Financing Facility for Aviation Safety' (9 December 2002), online: ICAO http://www.icao.int/icao/en/nr/pio200215.htm (date accessed: 24 December 2002).

The Canadian Transportation Agency, New release H100/99, 'Minister of Transport Issues Policy Framework for Restructuring of Airline Industry' (26 October 1999), online: Transport Canada

<http://www.tc.gc.ca/releases/nat/99_h100e.htm> (date accessed: 10 May 2001).

The Fraser Institute, Media Release, 'Avoiding the Maple Syrup Solution: Restructuring Canada's Airline Industry' (17 November 1999), online: the Fraser Institute http://www.fraserinstitute.ca/media/media_releases/1999/1999111 (date accessed: 8 September 2001).

Swissair Group, New Release 20/01/DC, 'Swissair Group et le gouvernement belge signent un accord sur Sabena' (17 July 2001), online: Swissair Group http://www.swissairgroup.com/apps/media/press/index.html/?period=archive& language=f#?period archive1language=f (date accessed: 27 September 2001).

## 1.7 Miscellaneous

CAPA minutes, *Memorandum of Understanding*, APA headquarters Fortworth, Texas, (9-10 February 2000), online: CAPA http://www.capapilots.org/Download%20Files/minutesfeb910.htm (date accessed: 14 May 2001).

Caribbean Alpa, *The problem with all Caribbean carriers is undercapitalization*, publication, online: Caribbean Alpa http://www.caribbeanalpa.com/discussion/posts/1546.html (date accessed: 14 May 2001).

'China opens up aviation industry' (8 August 2002) CNN, online CNN http://www.cnn.com/2002/WORLD/asiapcf/east/08/08/china.aviation/ (date accessed: 20 September 2002).

Ross, T.W., Stanbury, W.T., 'Avoiding the Maple Syrup Solution: Comments on the restructuring of Canada's Airline Industry' publication (1999), online: the Fraser Institute http://www.fraserinstitute.ca/publications/pps/32/ (date accessed: 8 September 2001).

Gavlak, D., 'Airline Industry Suffered Massive Losses After September 11' voanews (11 September 2002), online: voanews http://www.voanews.com/PrintArticle.cfm?objectID=5E25A889-57BA-49EF-A89D2 (date accessed: 19 September 2002).

## 1.8 Websites Visited

| | |
|---|---|
| Air France | http://www.airfrance.fr |
| Australian Commonwealth Department of Transport and Regional Services | http://www.dotrs.gov.au |
| Canadian Parliament | http://www.parl.gc.ca |
| Canadian Transportation Agency | http://www.cta-otc.gc.ca |
| Caribbean Air Line's Pilots Association | http://www.caribbeanalpa.com |
| CNN | http://www.cnn.com |
| Coalition of Airline Pilots Associations | http://www.capapilots.org |
| Council of Canadians | http://www.canadians.org |
| Department of Transportation | http://www.dot.gov |

| | |
|---|---|
| DHL | http://www.dhl-usa.com |
| European Union | http://www.europa.eu.int |
| Financial Times | http://www.ft.com |
| Fraser Institute | http://www.fraserinstitute.ca |
| International Civil Aviation Organization | http://www.icao.org |
| Le Figaro | http://www.lefigaro.fr |
| Le Monde | http://www.lemonde.fr |
| National Post | http://www.nationalpost.com |
| Swissair Group | http://www.swissairgroup.com |
| The Age | http://www.theage.com.au |
| The Scotsman | http://www.thescotsman.co.uk |
| Transport Canada | http://www.tc.gc.ca |
| Voice of America | http://www.voanews.com |
| Washington Post | http://www.washingtonpost.com |

# Index